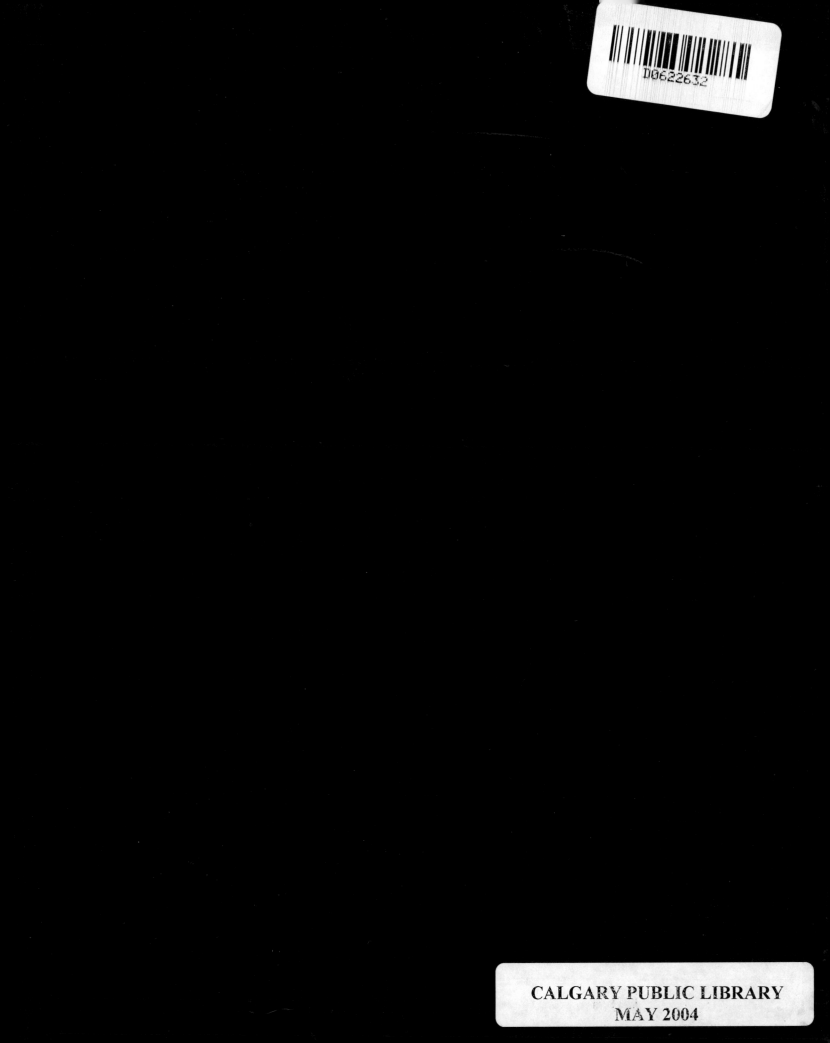

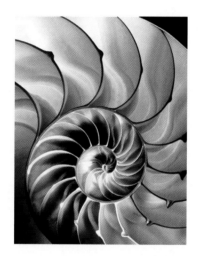

The New
Everyday Science

From the Big Bang to the human genome . . . and everything in between

Explained

By Curt Suplee

NATIONAL GEOGRAPHIC
WASHINGTON, D.C.

Contents

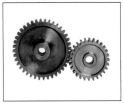
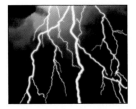
PRECEDING PAGES: The iridescent arc of a soap bubble conjures up mental images of spheres of much larger magnitude: the Earth and sun. Light playing across the bubble's surface breaks up into the colors of the rainbow.

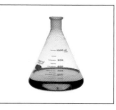
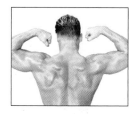

Living, breathing science

Understanding science—that is, how and why things work—needn't be dull or confusing. Believe it or not, this world we're in really does make sense.

t's one of those splendidly indolent mornings. You wake up and your world seems utterly still. For once, it appears, there's absolutely nothing going on.

But in fact, your surroundings are positively seething with activity. The light beams streaming through the window—produced 93 million miles away by nuclear fusion in the center of the sun—are slamming into your newspaper at nearly 186,000 miles per second. Thanks to chemical differences between newsprint and ink, the pages reflect most of this radiation, whereas the type absorbs it, producing patterns of alternating light and dark regions. Lenses in your eyes focus those patterns onto a curtain of nerve cells, and the signals are routed to various parts of your brain, where the peculiar squiggles are interpreted as sounds, letters, and words.

The cat meows. You hear it because atoms in the air collide with your eardrums at hundreds of miles per hour. Meanwhile, the air in the room writhes restlessly, rising as it's heated and falling as it cools, following the same cyclic convection pattern that, in the great outdoors, forms thunderstorms.

Meanwhile, your body cells are burning a simple sugar called glucose, keeping your temperature near 100° Fahrenheit. A powerful drug in your coffee—caffeine—is speeding to your brain, prompting release of a stimulant called norepinephrine, constricting blood vessels and making you more alert.

But not alert enough to detect the invisible torrents of radiation that are streaming through your room, or the trillions of subatomic particles called neutrinos that pierce your body every second, or even your dizzying velocity. Although you may think you're lying motionless, Earth's spin is actually moving you along at hundreds of miles per hour. You're also racing around the sun at about 18 miles per second, while the solar system is spinning through our galaxy at 140 miles per second—more than half a million miles an hour! All this, of course, as the universe continues expanding like a rapidly inflating balloon.

All these activities obey certain laws and rules, rules that in fact make our everyday world so reliably predictable. This book—now updated for the 21st century—is intended to help you look at the events of ordinary life in new ways, and thus better understand the hows and whys of Mother Nature's inexhaustibly diverse special effects. Topics are generally arranged from the simplest to the more involved, you can open to any page you like and use the convenient cross-references on the right-hand edges to find related subjects. ■

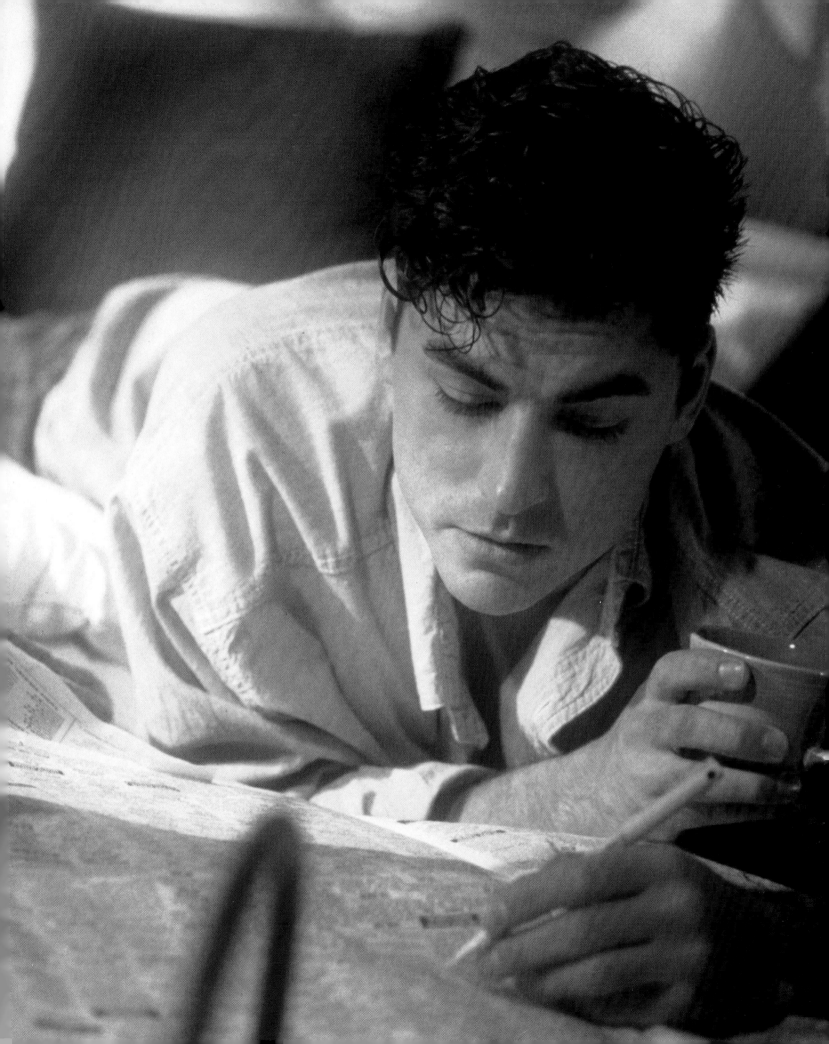

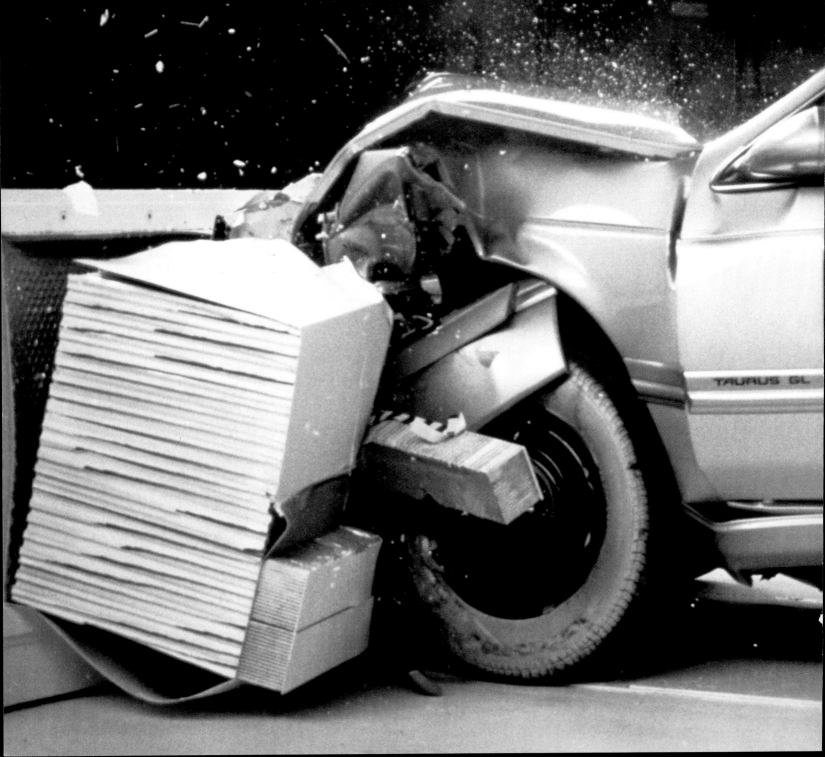

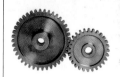

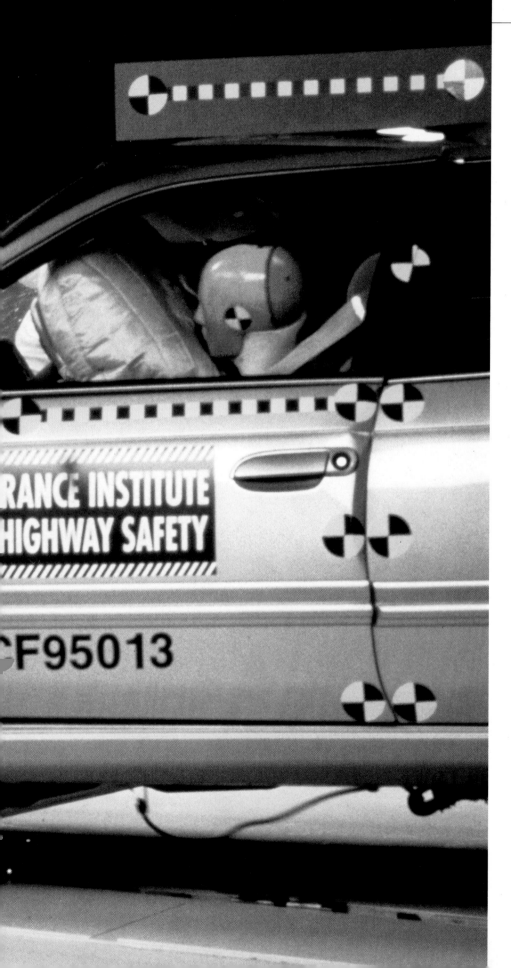

Matter
and
Motion

t's not surprising that human beings have spent thousands of years trying to understand exactly what happens when push comes to shove. After all, events have the most memorable impacts on our lives, and events are just collections of different things in motion. Physics—the science of matter and energy—determines the design of our automobiles, inside as well as out. It dictates the way our homes are built, how the pipes and wiring are laid out, and the design of the tools we use. It also affects the way your body parts are connected and how they move. The same rules that govern the stately circuit of the planets around the sun also bind us to Earth—and enable us, at times, to break those bonds and soar.

It's all downhill from here

The nature of the world around us has a lot to do with the two laws of thermodynamics—and with the curious concept of entropy.

Mother Nature is an infallible accountant, and impartial to a fault. Meticulous accounting is essential because energy can neither be created nor destroyed, though it readily changes forms. This principle, basic to the first law of thermodynamics, guarantees that there is always exactly as much energy around after something happens as there was before.

Impartiality is ensured by the second law, which decrees that everything generally tends toward the lowest energy state possible and the most equal distribution of resources. That's why the cream you pour into your coffee blends itself evenly throughout the cup, even if you don't stir it.

Natural events always proceed toward states of greater disorder. Knock over a glass vase and it shatters. Stomp around long enough and your shoelaces come untied. We never encounter the reverse sequences: Shoelaces don't spontaneously retie themselves. This one-way aspect of nature is called entropy, and in the long run it always increases. It's one law we just can't break. ∎

Why water falls ▶
This intrepid kayaker is demonstrating two key principles. One is the way energy changes form: In this case, the potential energy that he and his kayak possess by virtue of their height is being transformed into kinetic energy of motion. He loses elevation, but gains speed. And his downward trajectory illustrates that everything is seeking its lowest energy state. Either way, he's going with the flow.

◀ It's a messy, messy world
All systems in nature tend to proceed toward states of greater and greater disorder. That's why a teenager's room, left to itself, always gets messier, never neater. Fortunately, entropy can be reversed on the local level. But doing so entails expending energy.

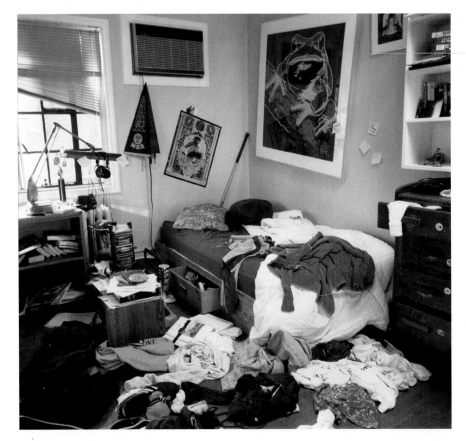

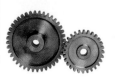

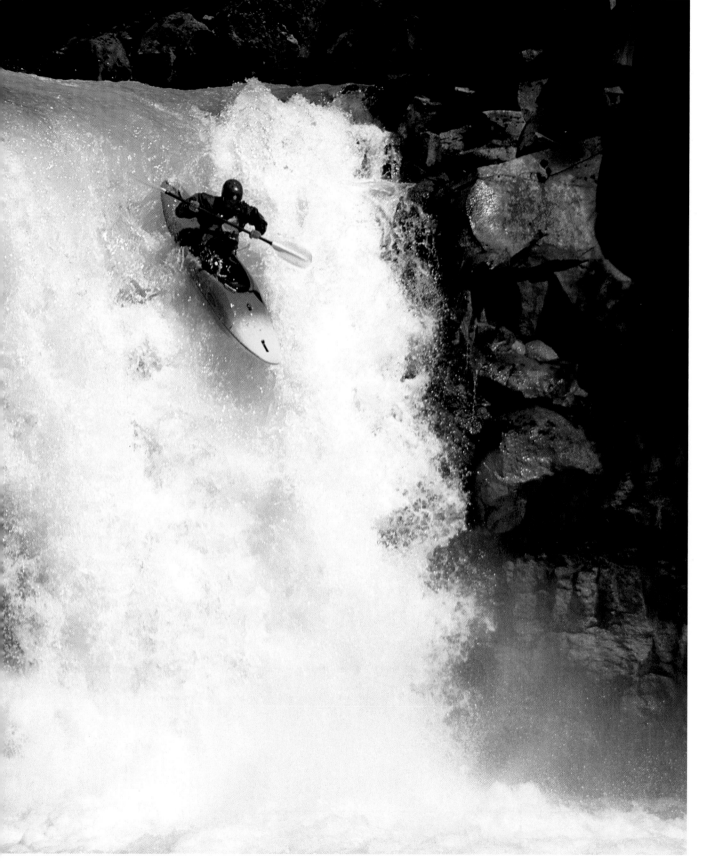

Orders of magnitude

Nature's laws operate over a mind-boggling range of scales, from the supergalactic to the subatomic. That's one reason scientists use exponents to describe the long and short of things.

Some forces—such as gravity—act across the entire cosmos, shaping galaxies and gas clusters billions of trillions of miles apart. Others, such as the strong force that binds the nuclei of atoms together, play their roles in spaces too tiny for even the finest microscopes to view: millionths of billionths of inches. The metric system copes handily with this immense spread of scales. Unlike the system we use with its illogical jumble of multipliers (12 inches to a foot, 16 ounces to a pound, and 5,280 feet to a mile), metric units are neatly arranged in powers of ten: Ten millimeters to a centimeter, a hundred centimeters to a meter, a thousand meters to a kilometer, and so forth. A meter, by the way, was once defined as one ten-millionth of the distance from the Equator to the North Pole, which comes to a tad over 39 inches.

Exponential terminology is simple and direct: "Kilo" signifies one thousand (or 10^3) of something, so a kilogram is a thousand grams, about 2.2 pounds. A million (10^6) watts is a megawatt; "giga" means a billion (10^9), while "tera" is a trillion (10^{12}). On the small side, "centi" means one hundredth (or 10^{-2}) of something, "milli" stands for one thousandth (or 10^{-3}), and "micro" means one millionth (or 10^{-6}). An average human hair is about a hundred micrometers (or microns) thick. "Nano" is the prefix for one billionth. Since a garden-variety atom is about one-tenth of a nanometer in diameter, "nanotechnology" refers to the process of manipulating energy and matter on the scale of individual atoms and molecules. ■

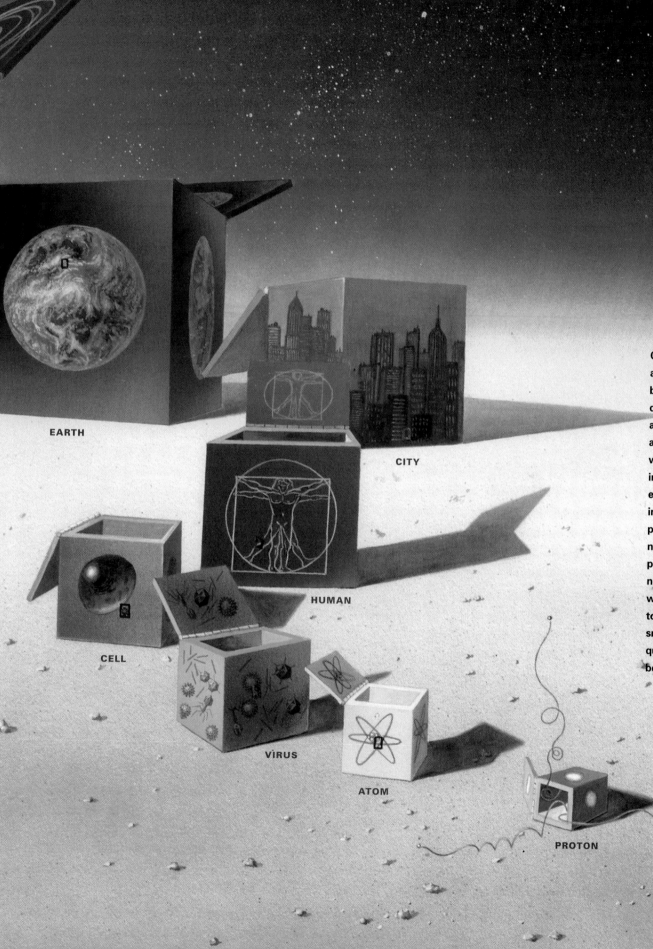

EARTH

CITY

HUMAN

CELL

VIRUS

ATOM

PROTON

QUARK

Orders of magnitude
are represented as
boxes of varying
dimensions. The
atom—named by
ancient Greeks for
what they considered
indivisible matter—
eventually was split
into much smaller
protons, electrons,
neutrons, and other
particles. Protons and
neutrons, in turn,
were revealed
to consist of even
smaller bits called
quarks. There may
be more surprises!

Getting started

Some basic definitions and rules of physics can help us rise above obstacles, keep our balance, and give us more time to hang around.

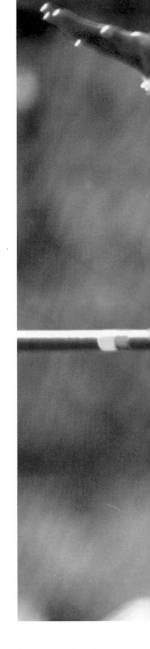

You may be pleased to know that, from the point of view of physics, your weight is comparatively trivial. That's because it varies according to where you find yourself. If you tip the Toledos at 180 pounds on Earth, you'd weigh only about 65 pounds on Mars, and a mere 30 pounds on the moon—even though it's the same you in all three locations. That's because weight is really a measure of how strongly two objects attract each other with the force of gravity. What doesn't change from place to place is your mass: the total bulk of atoms in your body. Even "weightless" astronauts in orbit have exactly as much mass as they had on the Earth's surface.

Mass determines not only your weight but also how hard it is to move you from one place to another. While mass comes in all shapes, under gravitational attraction it always behaves as if it's concentrated in a single point under which you can balance the load on your hand. For a juggler's club, it's the point around which the club rotates when you toss it. In an irregular and flexible form such as a human being, the center of gravity shifts constantly, depending on how the torso, arms, and legs are arranged. Sometimes, in fact, it can even be *outside* the body. ■

Mass and density ▶
The space that mass occupies is volume; the mass per unit of volume is density. The greater an object's mass, the more it resists being put in motion. That's why you'd rather kick a soccer ball than a bowling ball.

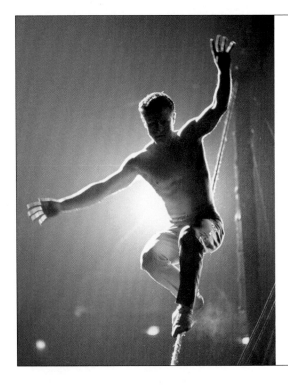

Center of gravity, center of mass

All of us exploit the concept of center of gravity, even though we may be unaware of the specifics involved. Children on a seesaw experimentally find their system's center of gravity by scooting up or back until everything balances on the fulcrum. A high jumper learns to curl his or her body around the bar (top right) rather than try to clear it upright. That's because the higher you raise your center of gravity, the more energy it takes. By rolling, you can get your entire body over the barrier—though your center of gravity actually may pass a few inches *beneath* it. Aerialists (left) keep their moving centers of gravity atop the rope by adjusting the position of arms, hips, and other body parts.

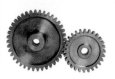

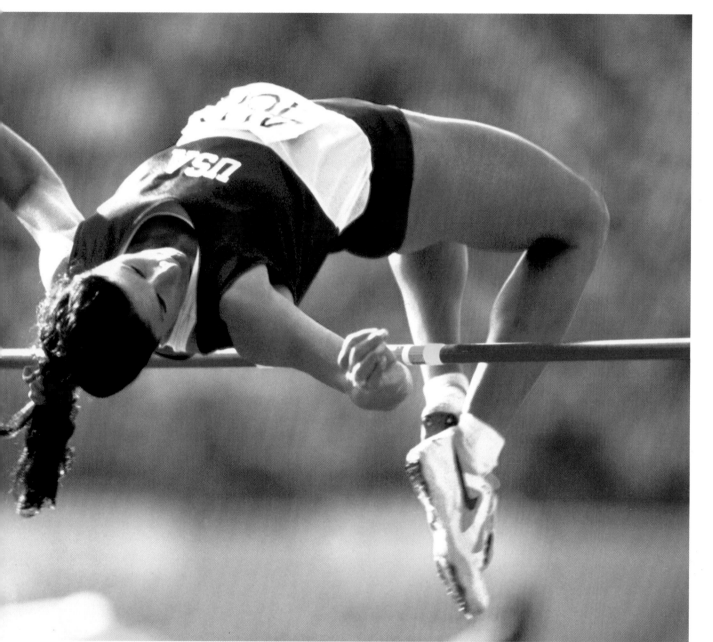

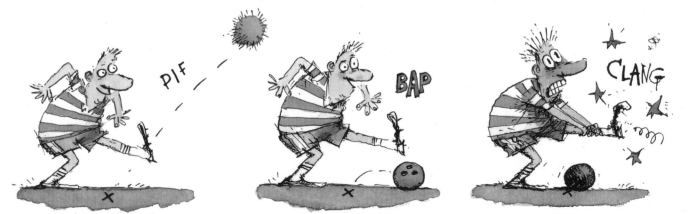

Newton I: inertia

Things that move obey certain laws. Three very important laws of motion were formulated by Isaac Newton (1642-1727).

ewton's first law of motion is deceptively simple: Objects at rest tend to remain at rest, while moving ones continue to move at uniform speed in a straight line—unless acted upon by an outside force or forces. This resistance to change is called inertia, and it explains a lot.

One is that when the car or subway train or airplane you're in begins to move, your body is "pushed back" against the seat. That is, your mass tends to remain at rest despite the forward-moving force of the vehicle, transferred to you through the seat.

Conversely, when your vehicle halts abruptly in a crash, your body tends to keep moving as it was. It would continue its natural graceful course straight through the windshield were it not for anti-inertial seat belts and air bags.

Another fundamental aspect of this law is that it confirms that the normal course for freely moving objects is a straight line. That explains why, when you whirl something around your head—such as a ball on a string—and then let it go, the ball flies straight. Good news for David. Bad news for Goliath. ∎

Turning it on ▶
Although we commonly use "acceleration" to mean speeding up, physicists define it as any change in speed or direction. So an object traveling in a curved path, such as this bicyclist, is constantly changing direction. Therefore it is accelerating, even though its speed (distance traveled per unit of time) may remain unchanged.

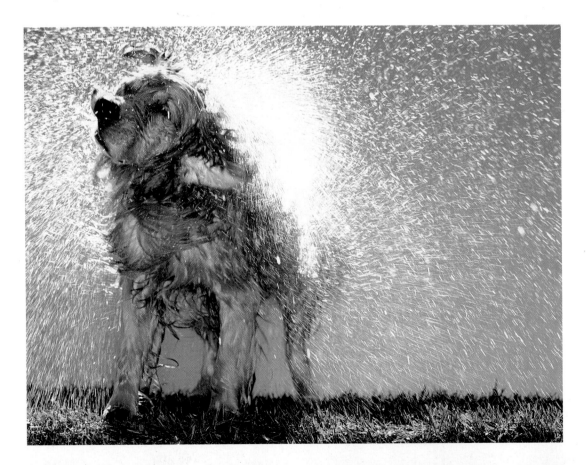

◀ Inertia's hair dryer
Few dogs perform well in physics classes. Yet they instinctively employ Newton's first law to shed water by setting their fur to flying in one direction, and then suddenly reversing course with a twitch of the muscles. The inertia of the water droplets exceeds the modest attractive force that binds them to the hairs, and they fly off—often onto the nearby owner.

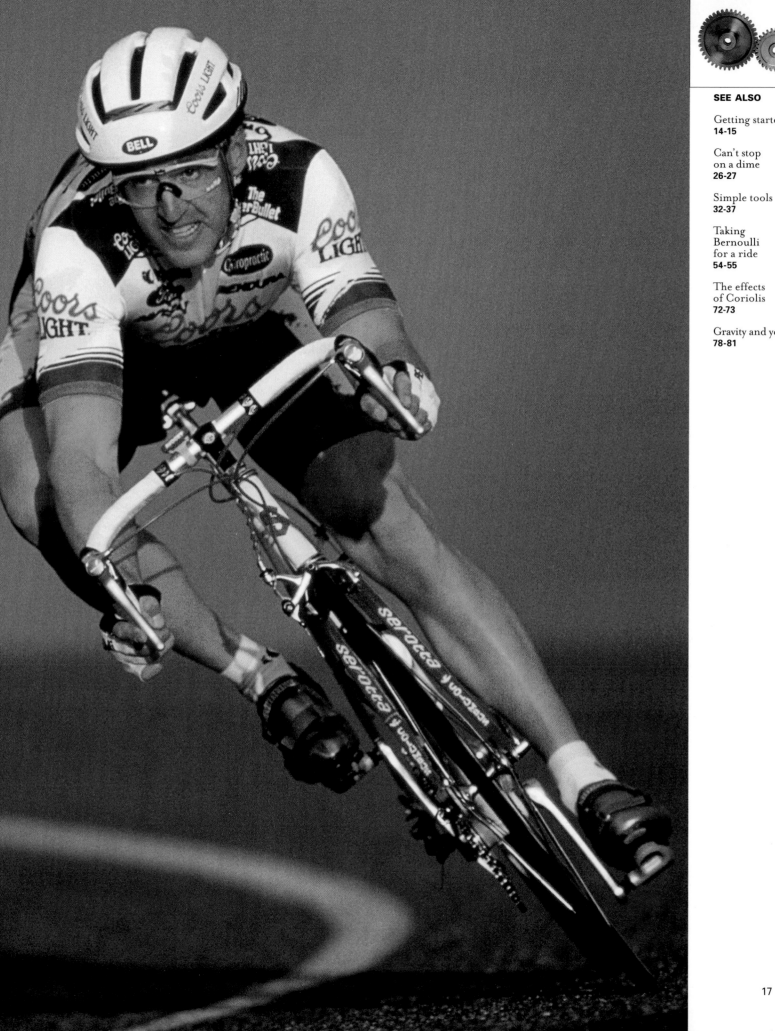

17

Newton II: F = MA

Newton's second law relates the amount of force needed to move an object to the object's mass and its acceleration.

Push a child in a swing, ride a planet around the sun, or shoot a game of pool, and you're using Newton's second law of motion, which states that whenever you want to change the speed or direction of something, you have to apply an appropriate force. The larger the mass or intended acceleration, the greater the necessary force. This law's straightforward formula allows engineers to calculate what's required to launch a jet fighter from an aircraft carrier, or how strong a seat belt must be to restrain, say, a 160-pound person when his car stops suddenly while traveling at 60 mph. It also is integral to planetary motion. Although at each instant inertia constantly urges a planet to fly off in a straight line, the awesome gravitational pull of the sun (which is roughly 330,000 times the mass of Earth) tugs it sunward during that same instant. The result: a stable, smoothly curved orbit.

So dependable is this law that apparent violations of it, observed in the 18th century, led astronomers to discover the unexpected planet Neptune. ∎

Planetary proof ▼
Turn it around, and the second law says that whenever acceleration occurs, a force is being applied. Consider a planet revolving around the sun. Because its curved trajectory is by definition acceleration, some force must be at work. That's gravity.

Work-free workout ▶
In science, "work" is defined in terms of forces acting over distances. An exertion that doesn't move something does no work. So no matter how much this weight lifter strains, he's only "working" when he actually pumps iron. However, a child pumping back and forth on a swing is doing solid physical work. Go figure.

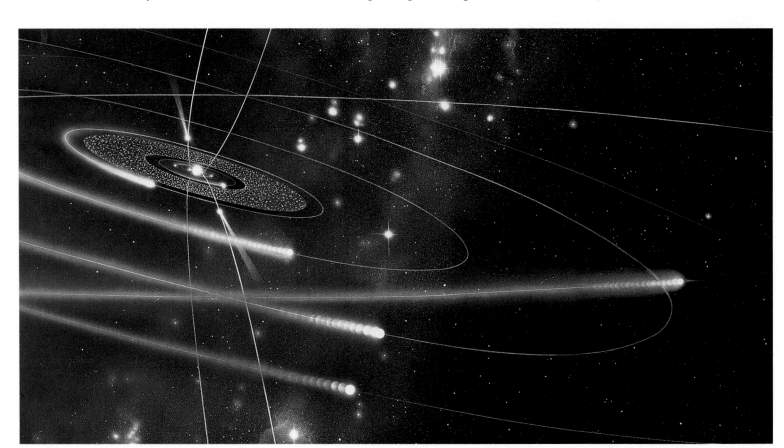

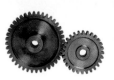

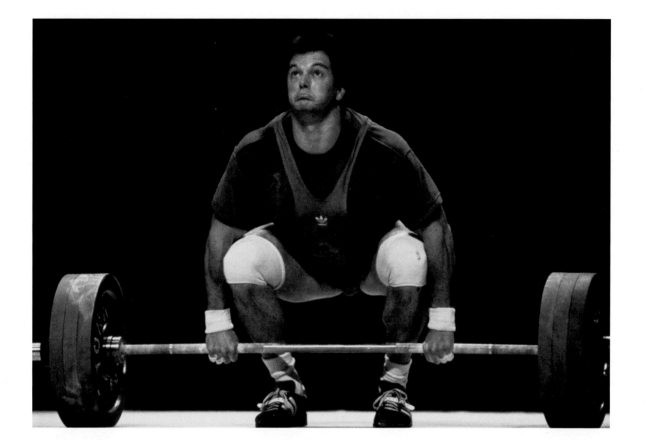

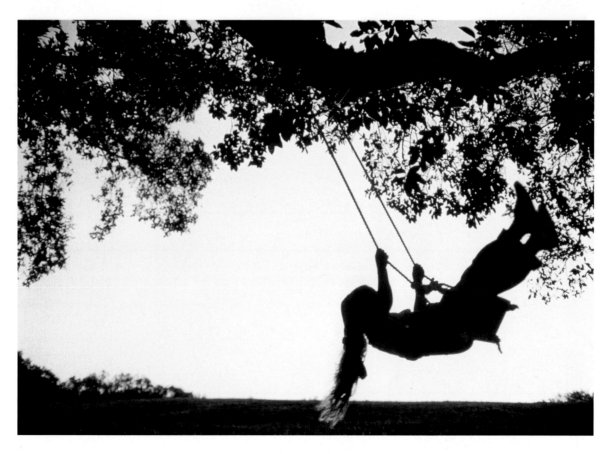

Newton III: equal and opposite

All objects, moving or stationary, obey Newton's third law of motion, which holds that in nature, unaccompanied forces do not exist.

You can't take a step without utilizing Newton's third law, which explains that for every action there is an equal and opposite reaction. Each time your foot pushes down on the ground, the ground pushes back with the same amount of force pointed in precisely the opposing direction. In fact, it's the force exerted by the ground (conveyed via friction to the sole of your shoe) that propels you along, not the force of your foot.

That seems odd, but it's easy to demonstrate. Stand on a nearly frictionless surface such as an oily floor, or put on a good pair of roller skates. Now take a step. Push as hard as you like, but the oil or skate bearings keep that force from being applied to the floor, so there is no opposite, reactive force exerted on your foot. The result: Your legs and feet may thrash about, but you don't move forward.

Ah, but suppose you've remembered to carry along a bag of baseballs at the same time that you're wearing those friction-free roller skates. *Now* you can move: Just start throwing the balls backward, over your shoulder. With each toss, you can exert a force on a ball. That ball's mass exerts an equal but opposite force on your hand, and you move forward a bit. Toss enough balls and you'll soon be moving at a nice clip. This action-reaction relationship accounts for the "kick" of a gun when it's fired: The forward motion of the bullet is balanced by a backward force on the shooter. The same principle makes rockets fly. ■

Goldilocks

As Goldilocks discovered during her brief housebreaking career, things are prone to fall apart when they are not strong enough to exert an equal but opposite counterforce to the stresses placed on them. In her case, that stress was caused by gravity pulling her toward the Earth with a force proportional to her mass. Baby Bear's chair just couldn't provide the equal but opposite resisting force. So it is that all successful designers and engineers—of chairs, parking garages, bridges, theaters, shelf systems, or what-have-you—must consider the implications of Newton's third law. That is, they must design their structures to meet or exceed anticipated loads.

Newton on the launch pad ▶
Action and reaction underlie the motion of jet and rocket engines. Both types of engines burn fuels that create expanding volumes of hot gases. Solid barriers within the engines confine the gases in such a way that they can escape only in one direction, like a bullet out of a rifle barrel. In effect, the engine walls press on the gas; the gas in turn pushes on those walls, shoving the plane or rocket in the desired direction.

When there's nothing to push back ▶
Not convinced that you need Newton to get by in this world? Try stepping from a dinghy to a dock. The force of your leg only serves to push the boat farther from the dock, and you get nowhere but wet.

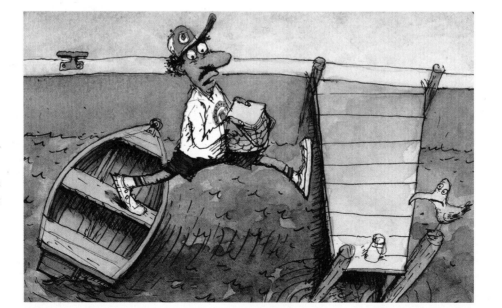

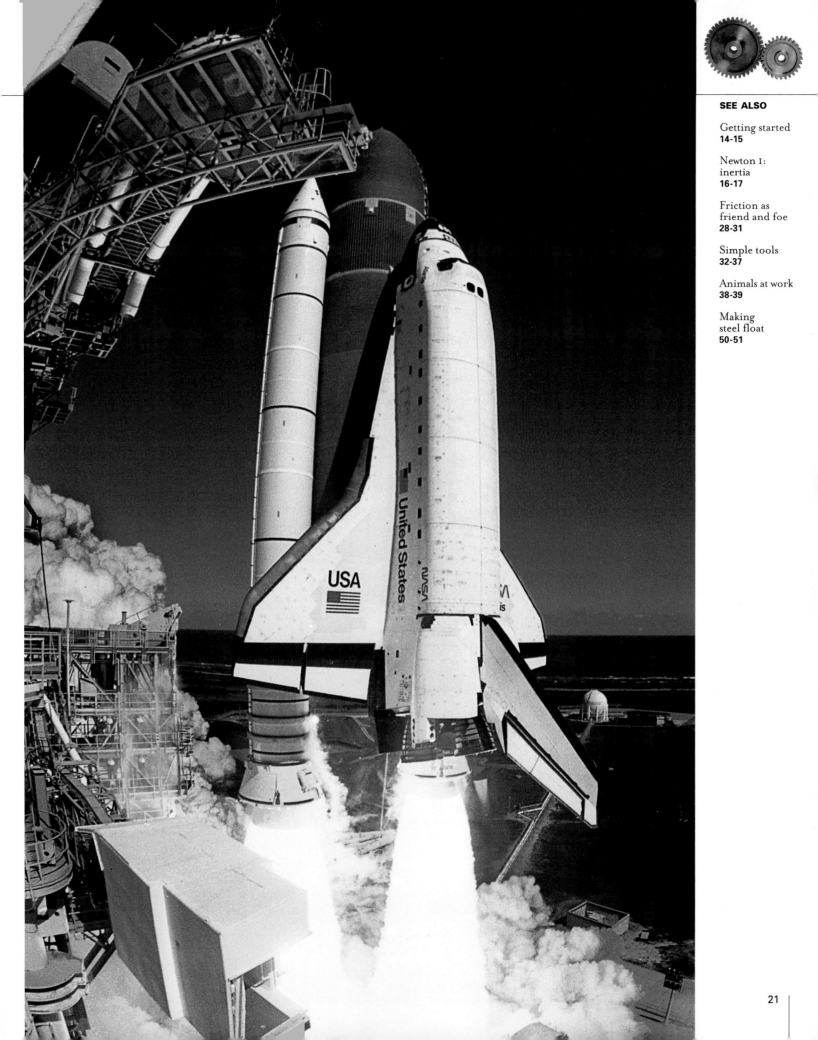

Neither created nor destroyed

Although energy can take many different forms—sometimes due to position, sometimes due to motion—the total energy in any system remains constant.

Energy is a little like art: You know it when you see it, but it's infuriatingly hard to define. In fact, the various forms in which it can occur—or into which it can be converted—are so diverse that science has given it only the broadest definition: Energy is what you need to do work, which is the ability to apply a force through a distance, that is, to move something. The amount of energy expended per unit of time is called power.

When energy consists of motion, it's called kinetic energy. Even a small, intrinsically harmless lump like a bullet can pick up lethal amounts of kinetic energy if its velocity becomes great enough.

But sometimes energy stems from the position or configuration of things, rather than from motion. That's potential energy. Potential energy has many faces: It exists in the taut curve of a drawn bow, in the pressure of compressed air, or in the tension on a highjumper's tendons and ligaments. It can be electrical, as in a battery, or magnetic. It also can be chemical, trapped in molecular bonds. High explosives detonate when the potential energy stored in very unstable chemical bonds is released. On a less violent scale, the same thing happens when your furnace produces heat from oil or gas. ∎

Ups and downs of energy conversion ▶
In the course of a day's skiing, you trade energy forms back and forth dozens of times. Electrical energy is harnessed in hauling you up the ski lift, which increases your potential energy of position. On the way down, you exchange that potential energy for kinetic energy of motion. And you stop or turn by converting kinetic energy to mechanical pressure that compresses (and warms!) the snow.

Conservation of energy

Next time the neighborhood kid puts a ball through your window, console yourself by pondering the energy conversion involved. Some of the ball's kinetic energy was used to overcome the attraction between molecules in the glass and send the fragments flying. Some was converted into molecular vibrations as heat. Some was transformed into acoustic energy—sound. Although you've lost a window, no energy was lost—it just changed form. Converting different forms of energy into one another is like changing dollars into British pounds or Swiss francs: it's fairly easy, but nature imposes a service charge for each transaction in the form of friction or other losses of "useful" energy. Thus, we never get as much out of any process as we put into it.

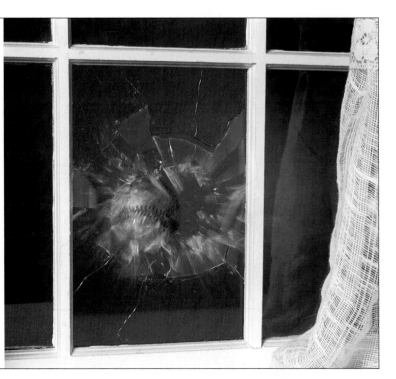

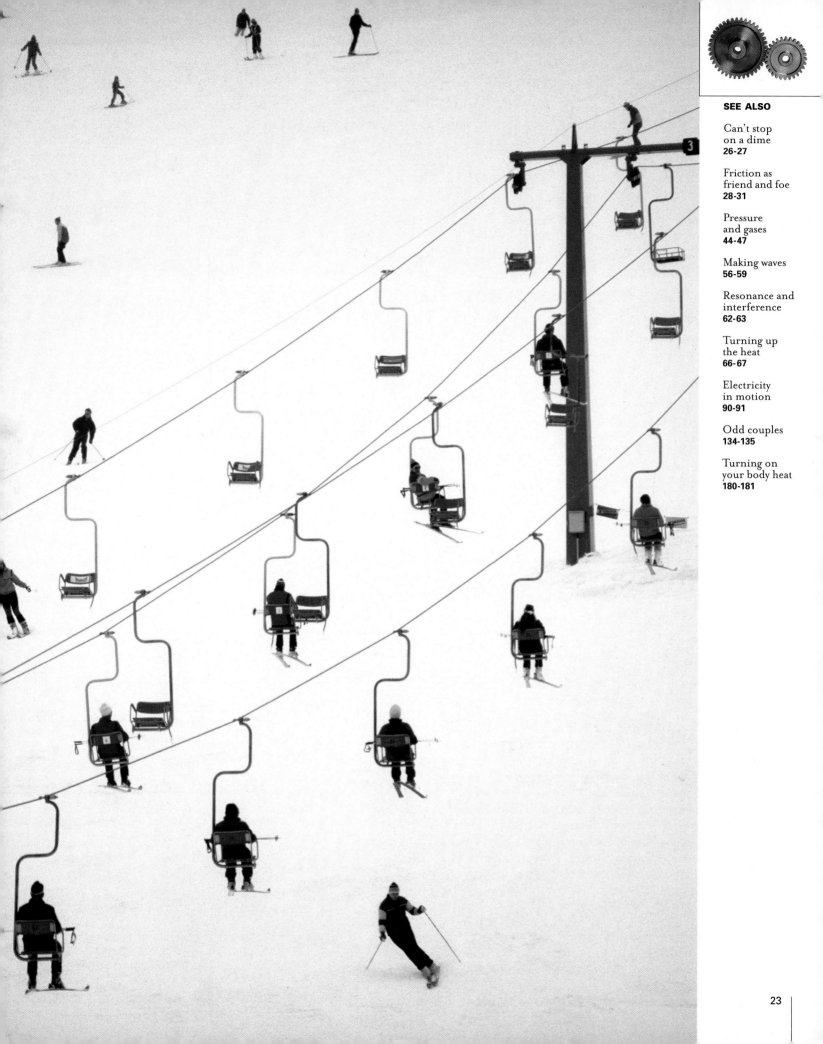

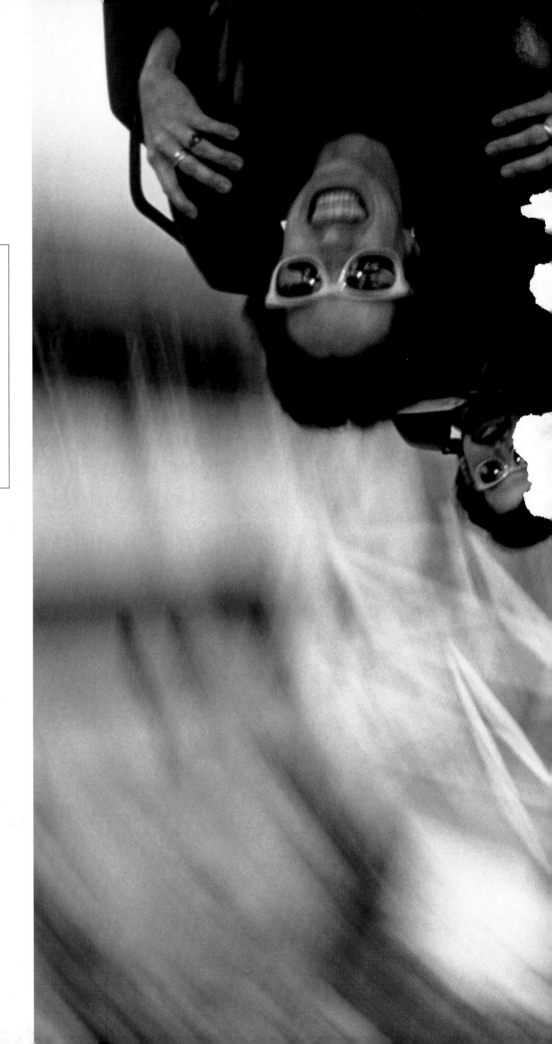

Energy to spare

It's not at all certain that these folks are thinking straight. But if they were, they'd probably realize that the roller coaster they're riding works simply by converting potential energy—obtained by hauling the cars to the start, the physical high point of the track—into kinetic energy, expressed in the speed of the coaster as it plunges downhill. How quickly the cars accelerate due to gravity and their total mass (fully loaded) are factors engineers use to calculate how steeply to bank the turns.

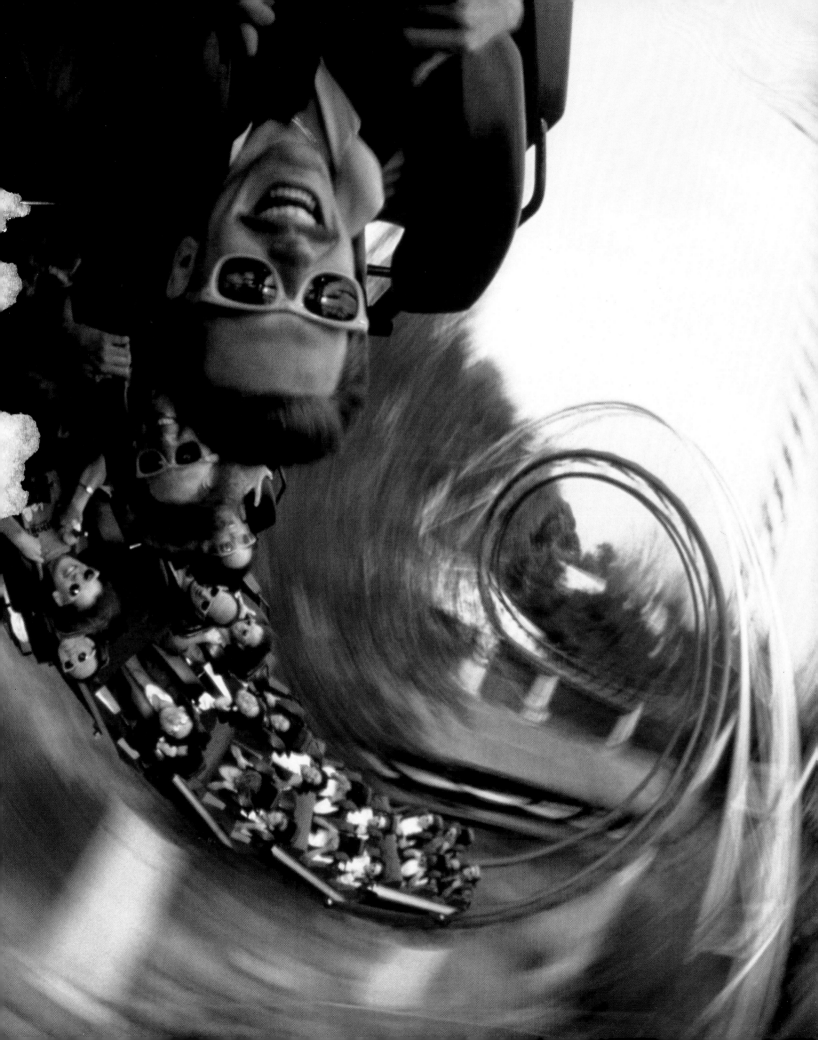

Can't stop on a dime

If you want to have a real impact on your world, you've got to build up momentum—the joint product of mass and velocity.

There are at least two reasons why life in the fast lane is dangerous: kinetic energy and momentum. Each is a different way of looking at the effects of velocity on mass, and just what happens when something hits something else. Kinetic energy increases in proportion to the mass of the moving object but varies with the square of the speed. Consequently, when you accelerate your car from 40 to 60 miles per hour, the energy you bring to a potential crash doesn't increase by 50 percent, it more than doubles. Think about that next time you put the pedal down.

Momentum, in contrast, is the product of mass times velocity, so it is directly proportional to both. A speeding bullet and a slow-rolling boulder can have the same momentum. Total momentum is always conserved. A rocket at rest on its launch pad has zero momentum. But as it starts to move, the principle of conservation demands that the rocket's momentum (big mass, low speed) is exactly equal to the momentum of the blasting gases of its exhaust (small mass, high speed) in the opposite direction. So, strange as it may seem, the total momentum of the entire system remains exactly zero! ∎

In the crunch ▼
Accident prone? Physics suggests that you need a big car, because in a collision the more massive object is more likely to keep moving in its original direction. And get one with built-in "crumple zones" that consume kinetic and mechanical energy as they deform.

Elastic vs. inelastic collisions ▶
Collisions in which kinetic energy is almost entirely conserved are called elastic. The degree of elasticity depends on the materials involved. When tennis racket meets ball, much of the kinetic energy is conserved. Not so for, say, tapioca.

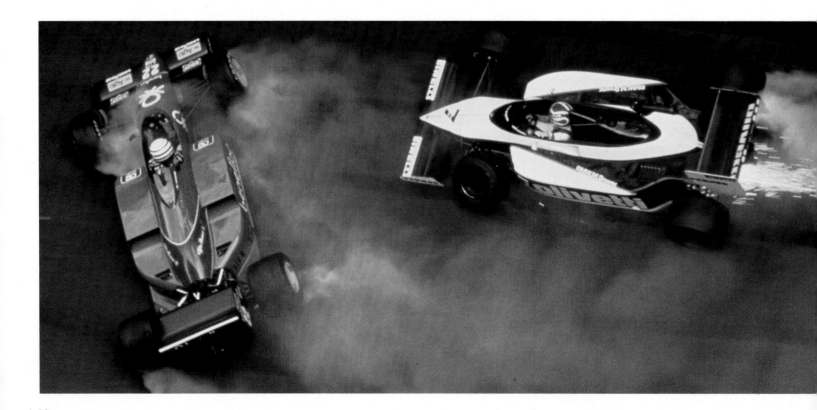

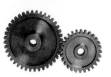

SEE ALSO

Getting started
14-15

Newton's laws
of motion
16-21

Resonance and
interference
62-63

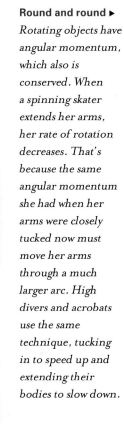

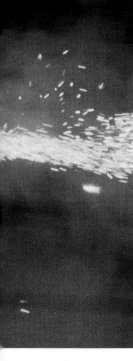

Round and round ▶

Rotating objects have angular momentum, which also is conserved. When a spinning skater extends her arms, her rate of rotation decreases. That's because the same angular momentum she had when her arms were closely tucked now must move her arms through a much larger arc. High divers and acrobats use the same technique, tucking in to speed up and extending their bodies to slow down.

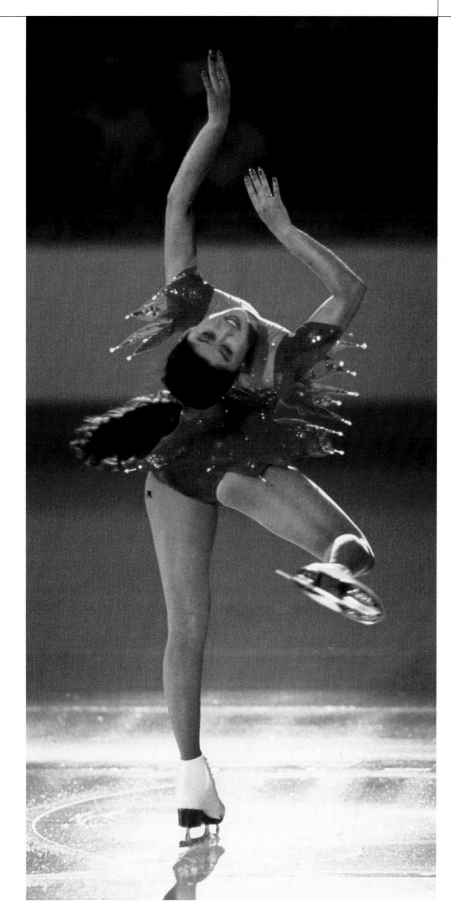

Friction as friend

Moving objects that come in contact with each other involve friction,
a resistive force with lots of everyday applications.

Materials to fit the feat ▶
Often it's to our advantage to maximize friction. Basketball or climbing shoes are engineered with flexible soles that increase contact area to stick to hardwood floors or hard-rock walls. Similarly, cleats expand surface area. Ordinary tires have grooves to keep water from squeezing in and dangerously reducing friction where the rubber meets the road. But at drag races and other dry-weather events, the tires are smooth for top total friction.

f you've ever scraped off a handful of hide or blistered your feet on a hike, you may justly regard friction as a major annoyance. But in fact, your life would be downright impossible without it. You couldn't walk, run, write with a pen or pencil, turn the page of this book, or even feed yourself. Many of mankind's niftiest inventions—including all sorts of athletic shoes—are specifically intended to increase friction in ways we find useful.

Friction is the tendency of two surfaces to stick together. They do this because their rough texture causes a raised part of one to get trapped in a low part of the other—as in sandpaper or asphalt roads—or because the atoms of the two materials naturally attract each other, or both. Any pair of surfaces has its own unique stickiness index, called the coefficient of friction. In fact, it has two coefficients: One for static friction, the force that must be overcome to set something in motion, and one for kinetic friction, the somewhat lower resistive force against something already moving. This fact explains why you should never spin your wheels on ice or snow: Friction is greatest just before the wheels start slipping.

For rubber on dry concrete, the static coefficient is 1.0 while the kinetic coefficient is 0.8. That means you would have to push with a force of a hundred pounds to get a hundred-pound rubber block moving, but with only eighty pounds to keep it sliding. In contrast, a hundred-pound metal block on a well-oiled metal floor (static coefficient 0.15, kinetic 0.07), would need a force of only fifteen pounds to get started and just seven pounds to keep going.

Violinists, cellists, and other string musicians turn friction into fugues. That's why they use rosin—a powdered form of sticky wood resins—on their bows. For similar reasons, many athletes grab for the rosin bag when they need to get a grip on objects from baseballs to balance beams. ∎

Fire through friction

The principal by-product of friction is heat. Sometimes that's desirable, as when you rub your hands to make them warm, or when you strike a match. In the latter case, friction generates just enough warmth to ignite a phosphorus-sulfur compound that in turn sets the rest of the match head aflame. Sometimes we need friction but not the heat, as in the case of automobile brakes. Braking systems slow wheels by pressing a heat-resistant pad against a metal surface—either a disc or a drum. But extreme heat can deform the pads and even begin to melt the metal. Consequently, brake drums and discs have vanes and other structures that quickly and harmlessly dissipate thermal energy into the air.

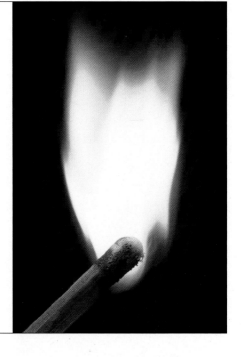

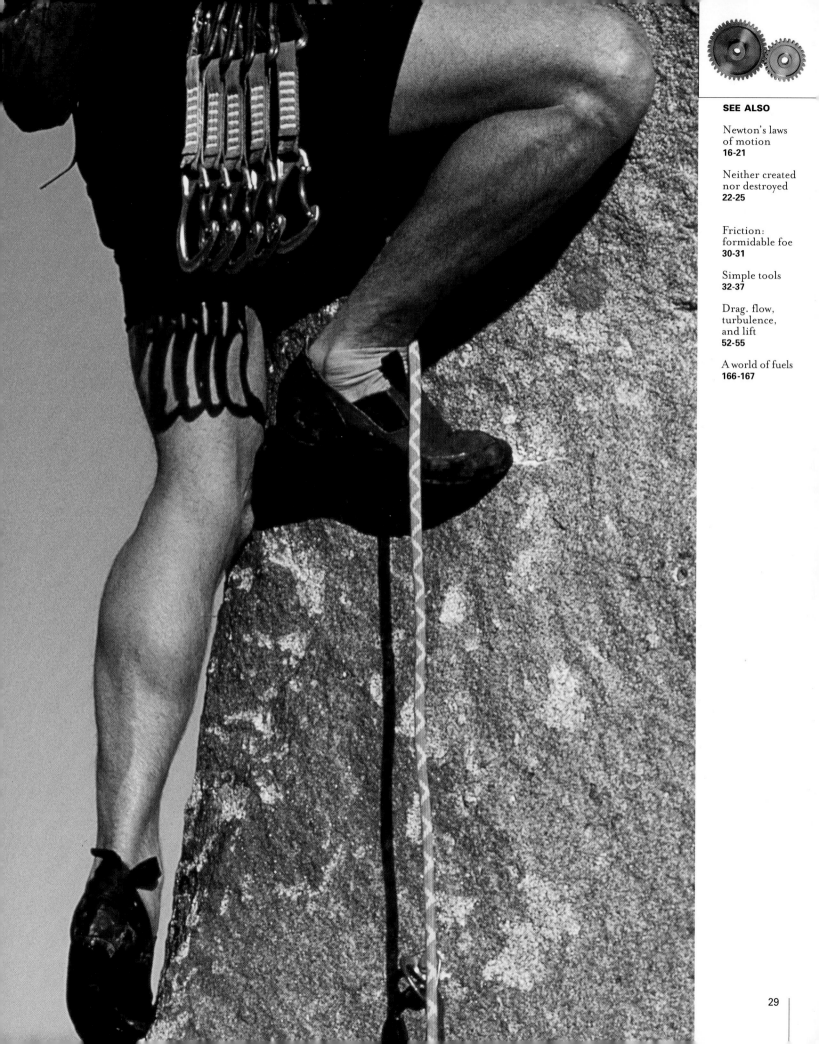

Friction: formidable foe

We may never totally eliminate it, but we've figured lots of ways to keep this resistive force to an absolute minimum.

Although we spend a great deal of effort finding ways to enhance friction, we probably consume even more in trying to reduce it, from spraying silicon aerosols on sticky window tracks to changing the oil in our cars or lubricating squeaky door hinges. And no wonder: Under the microscope, even the most highly polished metal bearings look as craggy as the Rocky Mountains. Two such objects moving against each other alternately stick and slide, generating heat and otherwise expending energy as useless noise and vibration. That's why devices such as these ball bearings (right)—which reduce the contact area between two metal surfaces to tiny points—pop up everywhere in our lives.

Another smooth move is to insert between the two surfaces a third substance that has a lower coefficient of friction. Many fluids qualify, including water in some circumstances. Unfortunately, it promotes rust and corrosion in iron and its alloys. But long ago, people discovered that oils from animal and vegetable fats would do the trick. Later they learned to utilize petroleum-based oils, which contain long, chainlike molecules that slide easily past one another. Oil mixed with a thickening agent makes grease, which is preferable in situations such as exposed joints on an automobile suspension. But solids such as graphite—a common, very soft form of carbon most familiar as the smeary "lead" in pencils—also can be used, especially in places where a fluid or paste might tend to trap abrasive grit. ■

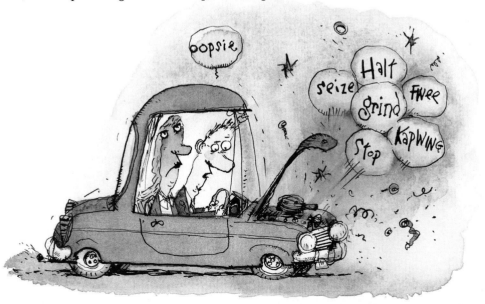

◄ Lifeblood of locomotion
Oil and other petroleum-based lubricants are not just slippery; they also prevent corrosion by keeping water and oxygen away from metal surfaces. Moreover, they can be modified chemically to increase their viscosity and to improve their ability to conduct heat away from machinery.

When sliding is a way of life ►
On the design side, engineers keep devising ever slipperier substances that can be worked into nautical hulls, ski bottoms, axles, and such. One popular synthetic is polytetrafluoroethylene, better known by its trade name of Teflon. On steel, it has a sliding coefficient of friction of 0.04. Pretty slick. It's used in nonstick cookware, thanks to a fairly high ignition temperature of almost 1,100°F—way beyond the demands of the average muffin.

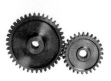

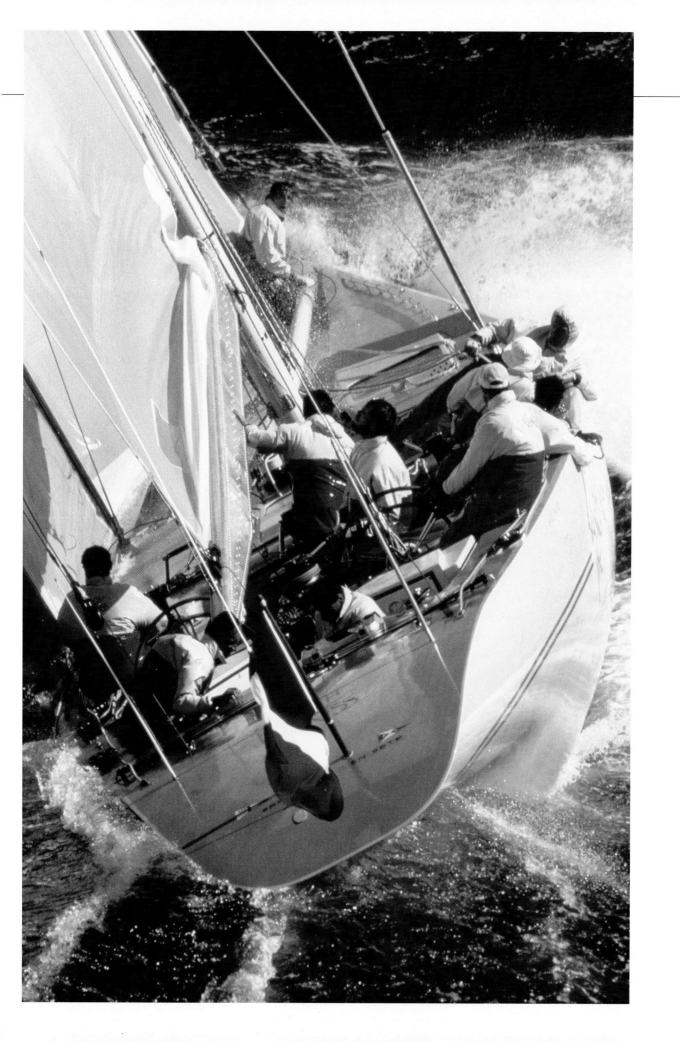

Looking for leverage

Even our most complex technological contraptions are collections of a few very simple but also very important machines. The lever is one of them.

That pain in your back is a reminder that, even in a well-oiled world, it still can be a hassle to move things around. So your aching ancestors invented machines to lighten the load. The most basic ones are levers, inclined planes, and hydraulic pressure systems. All of them adhere to one general principle, which sounds as if it were spouted by some guru of the 1960s: Work in equals work out. Since work is defined as a force exerted over a distance, machines can multiply either that force or that distance, but they can't do both at the same time.

All levers, including their circular derivatives such as wheels, gears, cranks, and pulleys, involve two lever arms—sections of the lever—and a fulcrum, or pivot point, around which the arms must move. Often the arms are on opposite sides of the fulcrum, as in a pry bar in which the bent area serves as the pivot, or an automobile jack, but they can also be on the same side, as in a wheelbarrow, a shovel, or a human forearm.

Any lever's mechanical advantage—that is, the degree to which it amplifies force—is determined by the ratio of its lever arms. If one is ten times longer than the other, you can use it to lift a 200-pound object by applying a force of only 20 pounds—but you'll have to push your lever arm ten times farther than the load moves. So powerful is this principle—and so well did Archimedes understand it—that the ancient Greek sage (below) boasted to his contemporaries, "Give me a place to stand, and I will move the Earth."

Pliers are a cunning combination of two levers that share the same fulcrum. Many tools, such as scissors and various cutters and snippers, add one or more inclined planes—in the form of beveled blades—to the double-lever arrangement. Wire cutters have long handles and short blades for maximum force at the cutting edge. Hedge trimmers and fabric scissors, which don't need much force, have long blades. ∎

More than a machine ▶

A corkscrew combines several physical machines. The handles are levers that end in rotary gears in which each tooth acts exactly as a conventional lever does. The gears connect to a toothed shaft that holds the corkscrew, which is an inclined plane reworked into spiral form. All together, these various machines enable you to draw the tightest cork with very little effort—because they are able to transform a modest force moving over a large distance into a large force moving over a relatively small distance.

Of wheels and wheelbases ▼

Leverage is what makes your car skid. Hit the brakes, and the body rocks forward, increasing the load on the front wheels and turning them into a fulcrum. That makes the car a lever, and it begins to turn. How much and how fast depends on the car's wheelbase— and the way in which weight is apportioned between the front and back of the car.

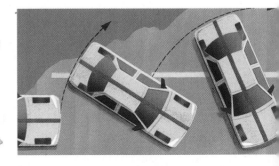

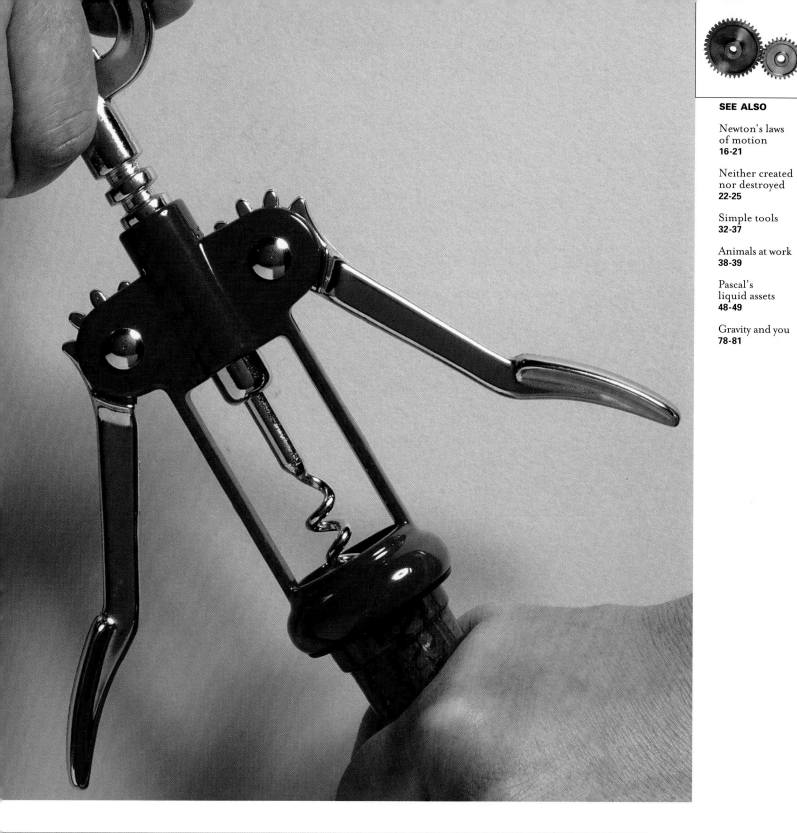

SEE ALSO

Newton's laws
of motion
16-21

Neither created
nor destroyed
22-25

Simple tools
32-37

Animals at work
38-39

Pascal's
liquid assets
48-49

Gravity and you
78-81

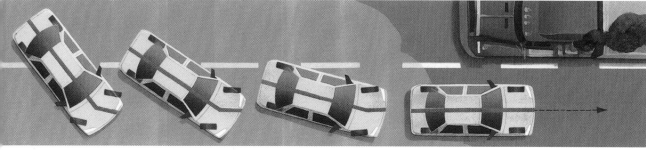

Wedges and wheels

Today's computerized world often ignores the simplest machines. But take them away, and we'd soon miss a lot more than crowbars and wedges.

A world without levers or inclined planes would have no doorknobs, no can openers, no stairways or ramps, no light switches, no pliers or knives or scissors or shovels, to name a very few of the conveniences we'd have to forego. And, of course, there would be no wheels. Wheels that are actually bound to solid axles, as in doorknobs and gears and rotary phone dials, are actually circular levers. Imagine putting a wrench on a nut and turning it through one-quarter of a revolution. That's precisely what happens when you turn your car's steering wheel 90 degrees or twist a doorknob. As with any lever, the mechanical advantage of a wheel depends on the comparative lengths of lever arms.

For example, when you pull a cart or push a wheelbarrow, you're actually exerting a force on the very bottom of the wheel that is equivalent to tugging on a wrench. The point on the wheel's rim that makes contact with the ground acts just like the tip of the free end of the wrench. The larger the diameter of the wheel compared to the diameter of the axle, the larger the system's ability to magnify force—just as a wrench or any other lever does. That's why mountain bikes in low gear use the largest rear sprocket to turn the back wheel. The mechanical trade-off is slower speed of rotation.

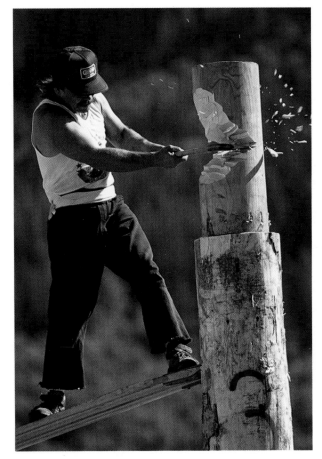

Wheels have other benefits as well. For one, rolling friction is usually much less than sliding friction. For another, frictional forces on an axle can be mitigated somewhat by using ball bearings or other slick devices that are easily protected by the axle's closed, cylindrical shape. ∎

Playing the angles ▶
The joy of an inclined plane—and its cousin the staircase—is that it allows you to stretch out the physical costs of reaching a vertical height. By following the diagonal slope of a ramp, a load travels farther to reach the same elevation, therefore requiring less force at any one time than if you had ascended the same distance by ladder. Stairs don't have the same mechanical advantage, but they give you a lift anyway.

◀ Putting edges on wedges
Most everyday cutting tools are inclined planes, simple wedge shapes with sharpened edges. A handsaw's teeth are nothing more than individual inclined planes in a row. Wedges were among our ancestors' first important tools, in the form of tapered flakes of stone and bone that provided early man with knife blades and projectile points.

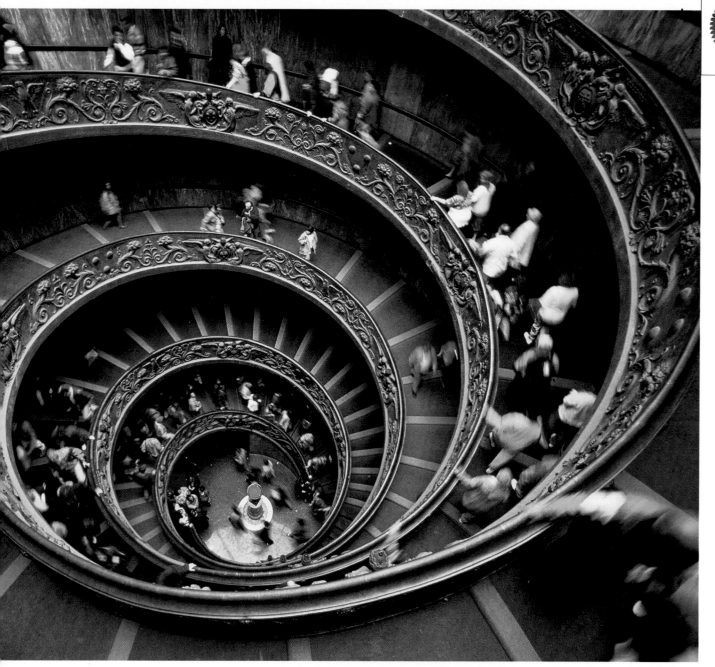

SEE ALSO

Getting started
14-15

Newton's laws
of motion
16-21

Friction as
friend and foe
28-31

Looking
for leverage
32-33

Pascal's
liquid assets
48-49

Gravity and you
78-81

The magic of the worm gear

Unlike most of the gears we encounter in daily life, the worm gear, whose "teeth" are a continuous spiral, resembles a screw in that it changes the direction of the applied force. Turn the worm gear pictured here, and it pushes the attached rotary gear on its axis—perpendicular to the force you exerted. Worm gears have been used in the steering systems of automobiles and other vehicles, converting the steering wheel's rotary motion into the horizontal, side-to-side movement of the front wheels.

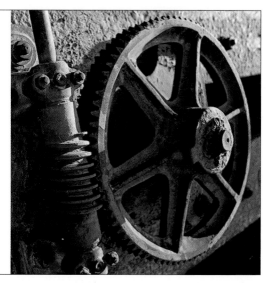

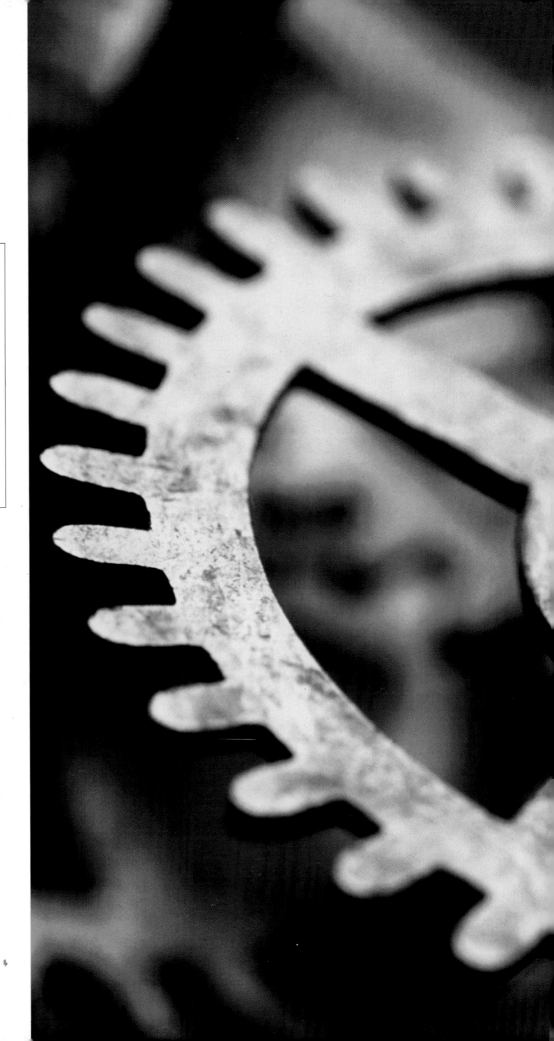

Levers with handles

Like simple levers, rotary gears can multiply either the amount of force applied or the distance traveled, but not both at once. When a small gear, such as the one in the center of this image, powers a larger one, such as the big wheel at right, then the driven gear rotates at a slower rate, but with increased force. That's what happens when you shift your transmission into low gear to get maximum power to the wheels. Alternatively, when a large gear drives a little one, the smaller turns much faster but less forcefully than its big brother.

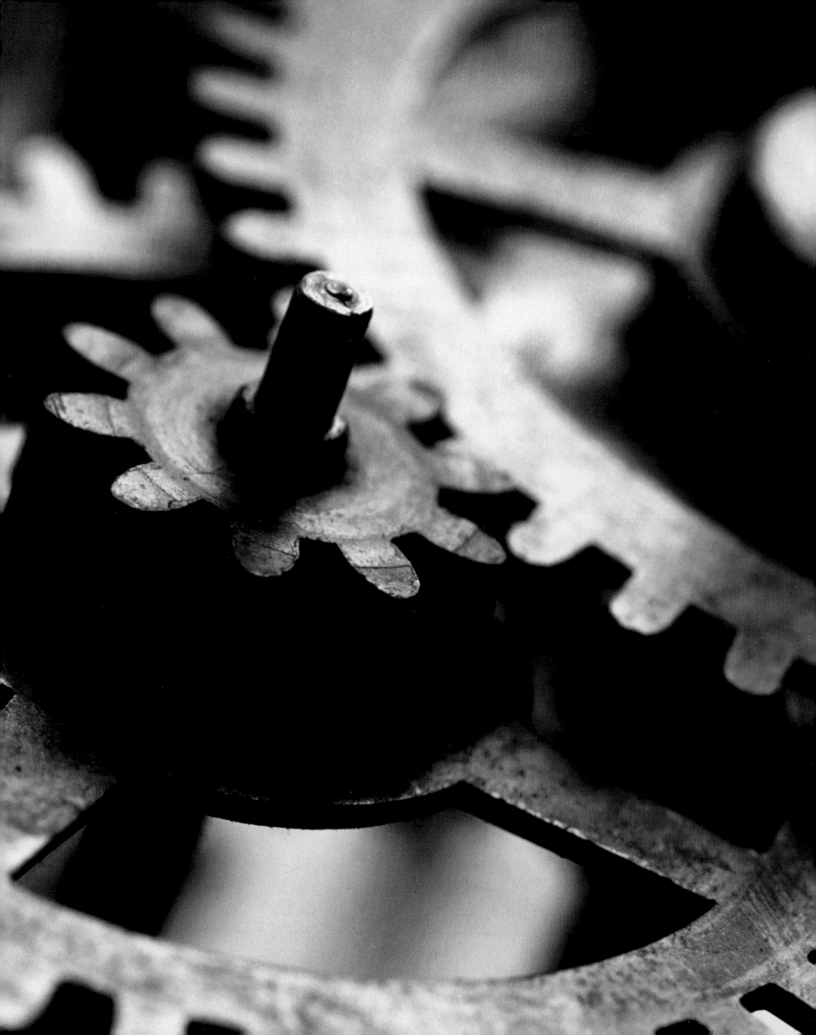

Animals at work

Tested both by time and evolution, the same simple machines that we call human inventions also abound in Mother Nature's varied workplace.

Keeping life on the move, nature equips all her creatures—people included—with numerous biological variations on the most basic of physical machines. Inclined planes serve nicely as teeth and claws. Worms and some other animals lacking stiff skeletons use hydraulic pressure to move, their muscles alternately lengthening and contracting fluid-filled body segments.

Animals with hard skeletons—internal or external—often rely on lever systems comprised of muscles, joints, and biochemical springs called tendons. Such systems appear in everything from cockroaches to gazelles to movie stars.

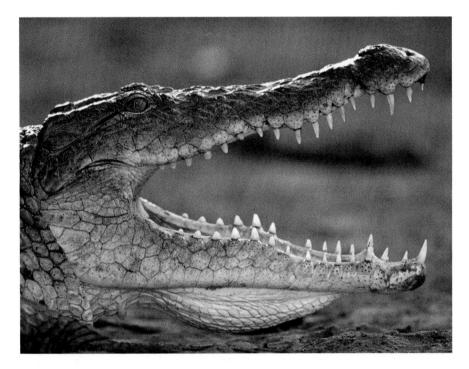

Running is a form of controlled falling. Every time your foot hits the ground, the force of your falling body is centered on a small part of the sole just behind the toes. Your ankle joint is the fulcrum of a lever that receives this huge force, and you'd simply collapse unless there were an equal opposing force. It could be provided entirely by muscle, but that would make running even more energy-expensive than it is. Instead, balance is provided largely by the pull of the Achilles tendon, which stretches by about 6 percent of its length on every running stride. That elastic stress—thousands of pounds per square inch—is released every time you spring up for the next step. ∎

Dwellers in liquid environments, ▶ *whether dolphins or their human companions, have no solid objects to push against. So most of them get around by exerting pressure against the resistance of their surrounding fluid. The dolphin's tail and the diver's rubber footwear do this by flexing at an angle, generating forward motion.*

◀ Nature: where flex is in flux
The jawbones of this crocodile constitute gigantic lever arms, studded with wedgelike teeth. The attached muscles exert large forces over short distances. In humans, the chief need often is to move relatively small weights over large distances—as in throwing a rock or running. So our muscles are fastened close to knee or elbow for less force of leverage, but wide range of motion.

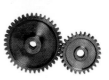

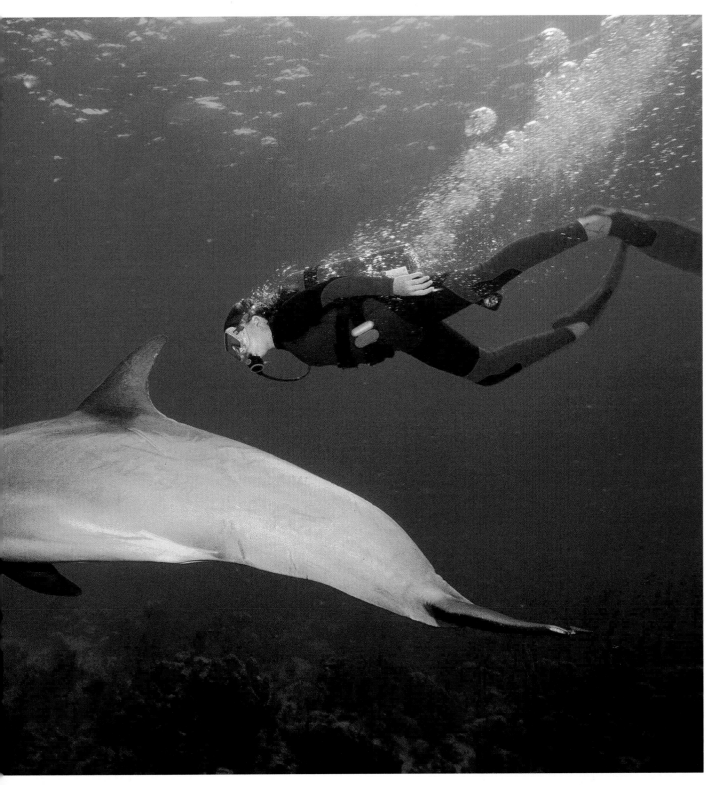

To stick or not to stick

While motion is essential to life, some things—living and otherwise—just hang around. Adhesives do the job by forcing surfaces to bind together.

Some natural bonding agents—sealing wax, pine tars, flours, and gluey proteins such as albumin and casein—have been used for centuries. In recent decades, scientists have devised a splendid variety of epoxies and other synthetic adhesives, some of which are far stronger and more resilient than the materials they join.

Adhesives work for two reasons. First, in the course of hardening, they form physical bridges between high and low irregularities in the surfaces involved. These mechanical attachments often are such that separating them actually requires you to rip apart the materials they're holding together. That's why nobody opens envelopes by separating the glued flaps.

Second, adhesives rely on the convenient fact that nature is inherently sticky. If you took two pieces of pure copper that were perfectly smooth and completely clean and brought them together in a vacuum, they wouldn't slide over each other at all. Instead, the copper atoms of one surface would bond immediately to those of the other, and the two pieces of metal would simply merge. ∎

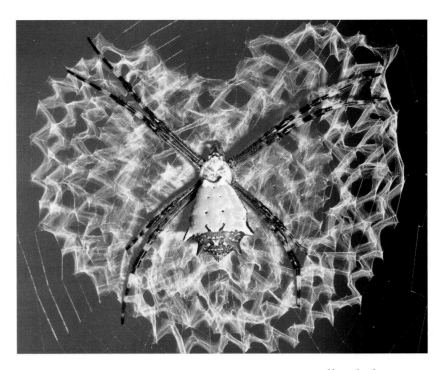

▲ **Hang in there**
Spider silk is a combination of proteins that, strand for strand, is stronger than steel. Spiders coat many of the strands with sticky droplets to catch insects. So why don't they get stuck in their own webs? Some experts believe that spiders step between droplets; others claim they cover their feet with oily liquid.

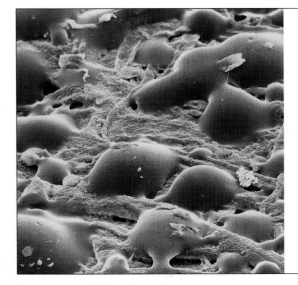

How much is enough?

Many of today's synthetic adhesives exploit both mechanical and inter-molecular bonding. So-called super glues contain relatively short molecular strands that quickly knit themselves into long chains known as polymers. Some are tough enough to be used in place of surgical sutures. Pressure-sensitive adhesives on strip bandages use various rubbery "elastomers" that are mercifully engineered to stick tighter to the plastic backing than to your skin. Similar elastomers make possible the on-again, off-again stickiness of Post-it notes (left).

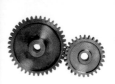

SEE ALSO

Friction as
friend and foe
28-31

Pascal's
liquid assets
48-49

**Why rain
comes in drops ▶**

*The droplet hanging
at the end of any
eyedropper owes
its existence to
cohesion, the
attraction between
like molecules.
Adhesion—the
mutual attraction of
unlike molecules—
also plays a role. Put a
glass straw in a cup of
water, and the surface
of the water inside
the straw takes on a
curved, concave
shape. That's because
adhesion between
water and glass
molecules is actually
stronger than
cohesion among
water molecules.
That's also why,
when a raindrop
trickles down your
windowpane, it leaves
a trail of tiny beads.*

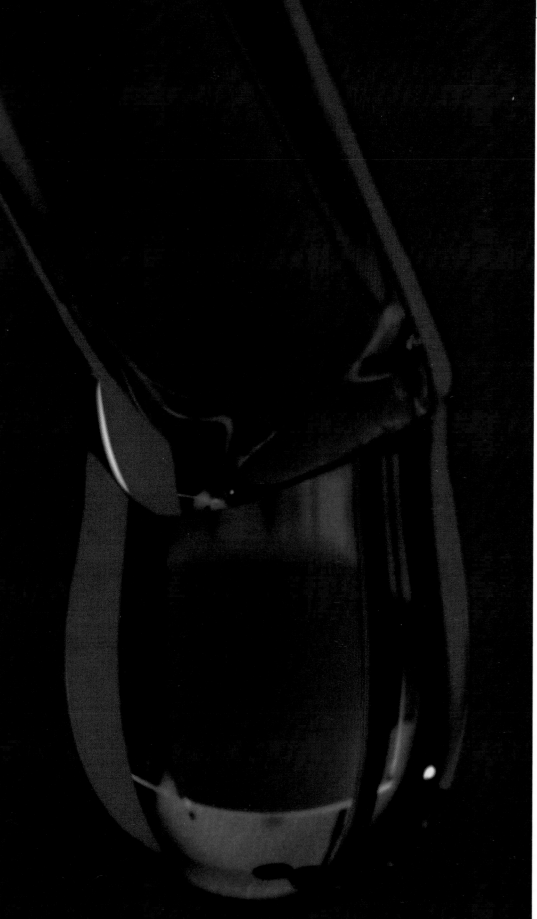

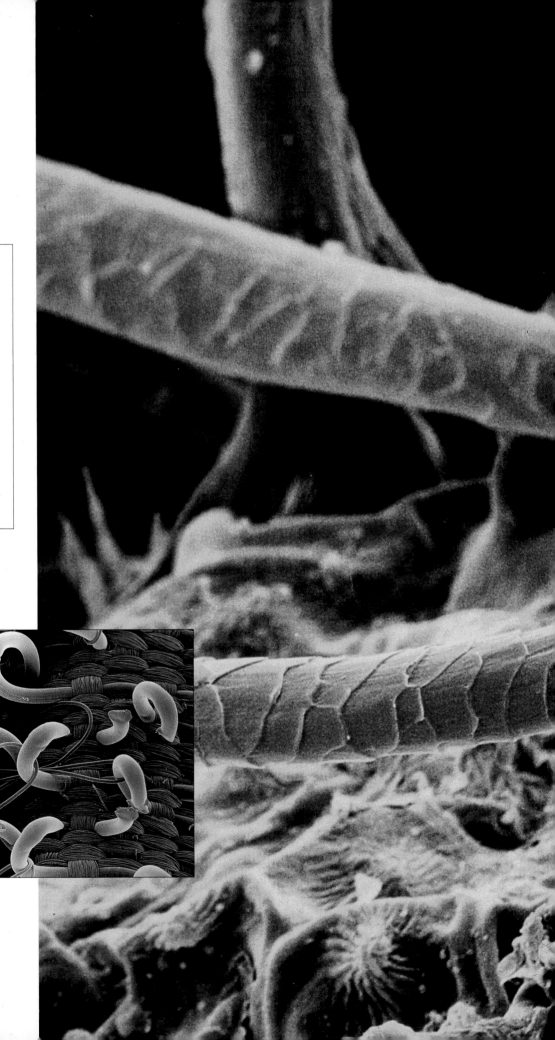

Getting around by sticking

Sometimes there's nothing more to adhesion than plain old mechanics. Each bristly bur of a goosegrass plant puts forth hundreds of tiny, hooklike snags (right) that readily attach to passing objects such as socks or furry animals. This system guarantees widespread dissemination of the seeds and, so far at least, biological success. Fascinated by the microstructure of burs, Swiss engineer Georges de Mestral invented a synthetic hook-and-eye material called Velcro (below), a contraction from the French words for "velvet" and "hook."

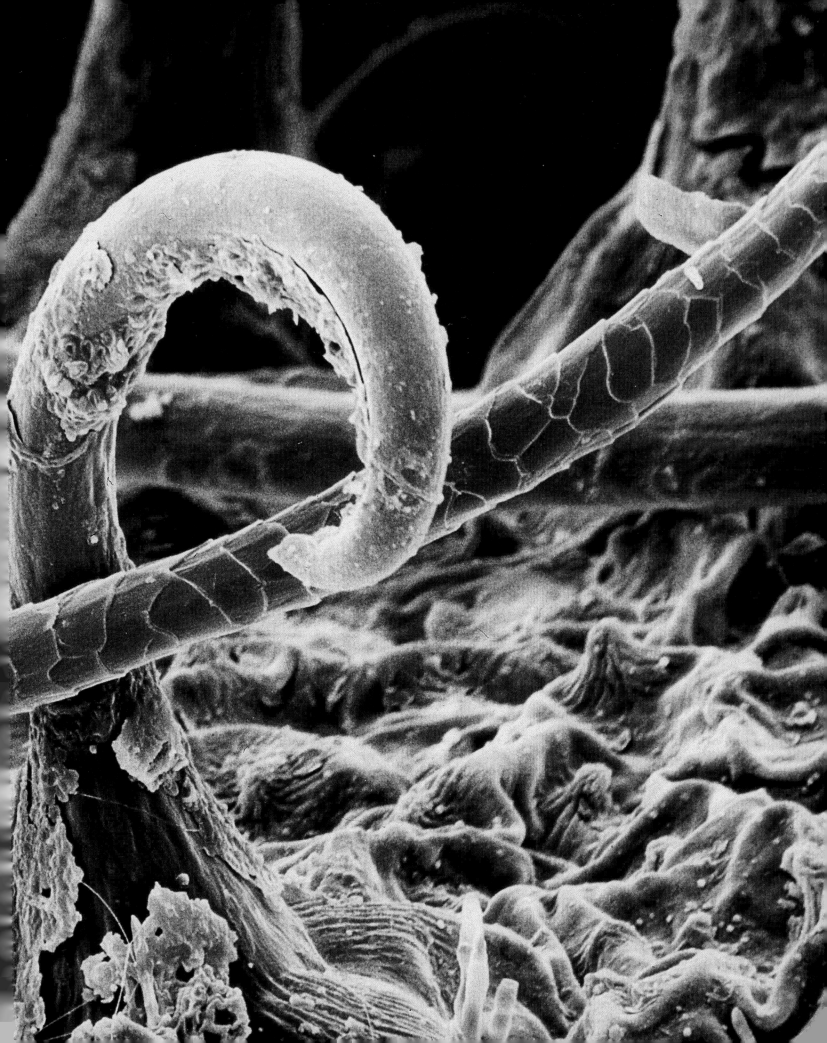

Pressure and gases

Unlike solids and liquids, gases possess neither definite shape nor definite volume. Left to themselves, they expand infinitely to fill any container.

hen you consider all the pressures of modern life, you probably don't think of the power of expanding gases, yet they account for much of civilization's more recent progress: Steam drove the 19th century, while internal combustion continues to propel the 21st. All gases obey simple but dynamic laws linking their temperature, pressure, and volume. Reduce the volume while keeping temperature constant, and the pressure rises in direct but inverse proportion—as occurs inside a bicycle pump. Increase temperature while keeping volume constant—as in a pressure cooker—and the pressure rises proportionally. By using energy to force changes in these variables, you can even turn hot air into work. Your car's engine does that all the time. On its intake stroke, each piston descends within its cylinder, drawing in a vaporous mix of air and gasoline. Then it rises, reducing the volume and increasing both pressure and temperature. At the top of the stroke, a spark plug ignites this explosive mixture, releasing stored chemical energy that boosts the heat and pressure even more. The resulting pressure drives the piston down again in the power stroke, and that energy is transferred via the crankshaft and other mechanical connectors to your wheels. ■

Getting that kick ▶
The amount of gas that can dissolve in a liquid varies with temperature and pressure. In champagne, fermentation converts sugar to alcohol and carbon dioxide is produced, raising the pressure inside the bottle enough to force much of the gas into the developing wine. Pop the cork, and you suddenly reduce the pressure to normal atmospheric levels—and the trapped gas just can't wait to get out.

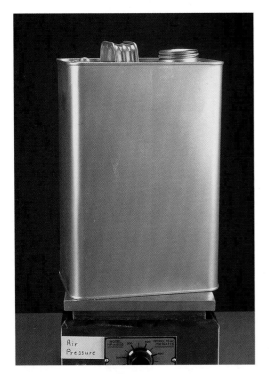
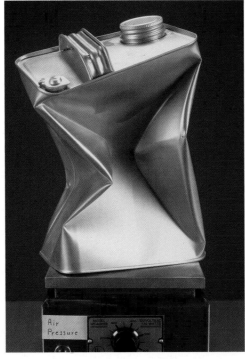

◀ The big squeeze
Amaze your friends with this never-fail party trick. Heat a can, raising the temperature (and hence the pressure) of the air molecules inside. Then seal the can and let it cool. As the temperature drops, so does the pressure and average velocity of the ricocheting air molecules. Result: implosive collapse.

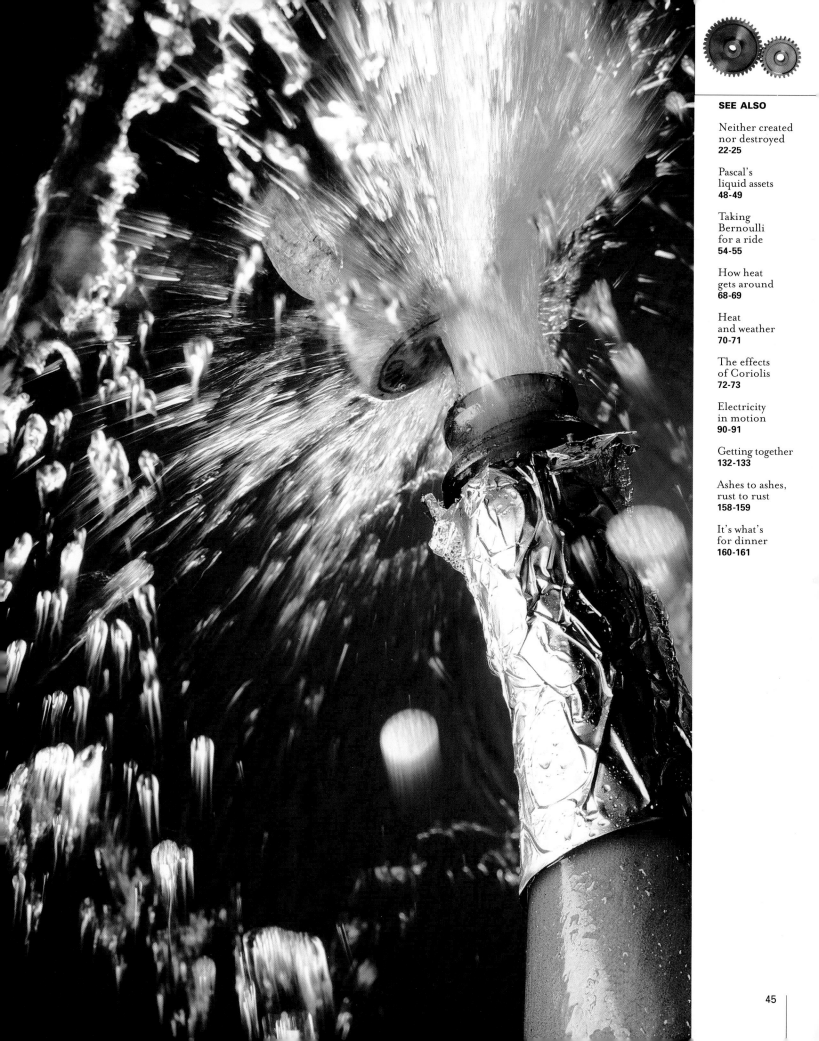

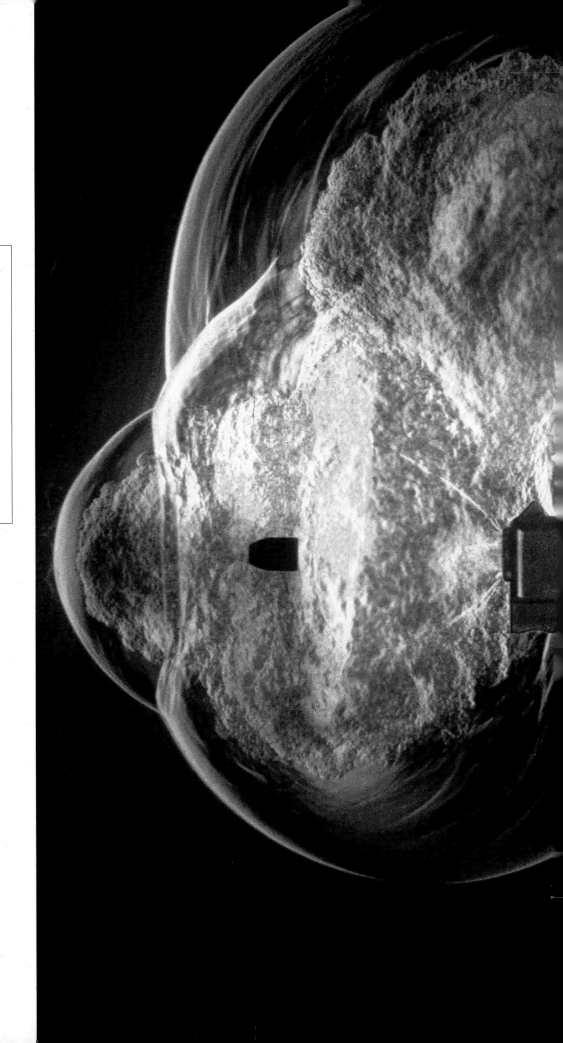

Speeding bullet

Gunpowder doesn't get the lead out; rapidly expanding gases do. In the fraction of a second it takes a bullet's explosive charge to detonate and the bullet to exit the barrel, supersonic shock waves of high-density gas spawned by the explosion have already blasted into the surrounding air. A special strobe with a duration of only 1/500,000 of a second helped freeze the action in this remarkable photograph. Billows above the barrel result from gases escaping through exhaust ports along the top of the gun, which help offset the upward component of recoil.

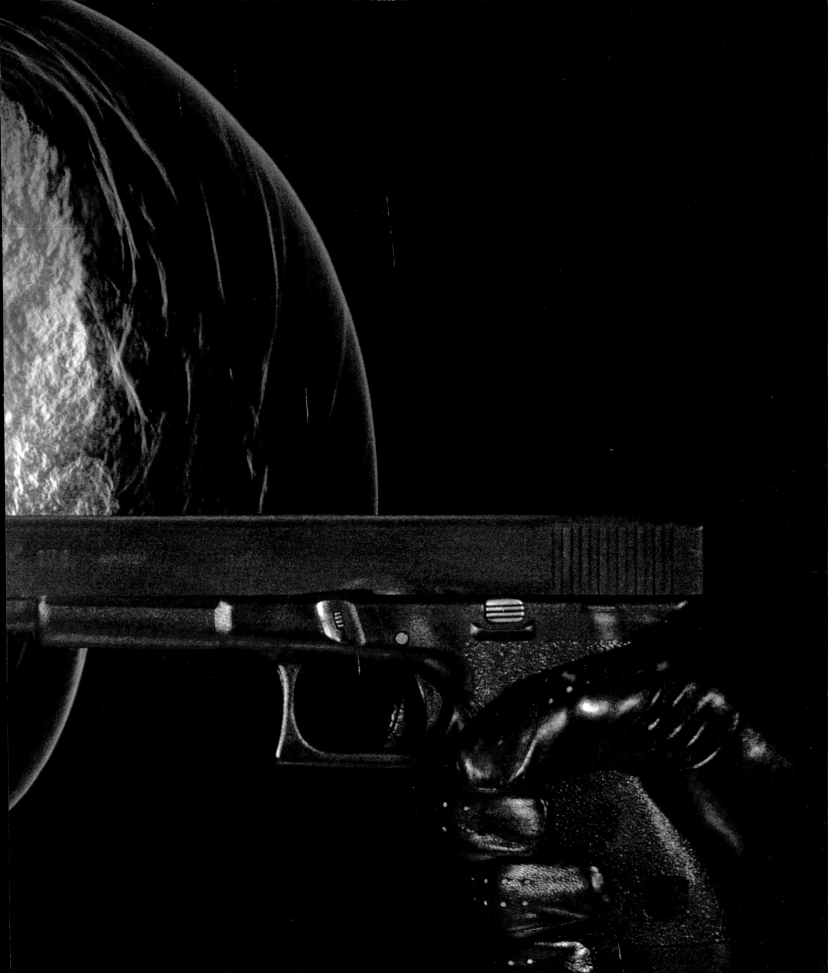

Pascal's liquid assets

Gases and liquids are fluids—states of matter that flow and follow the shapes of their containers. But they don't behave the same way under pressure.

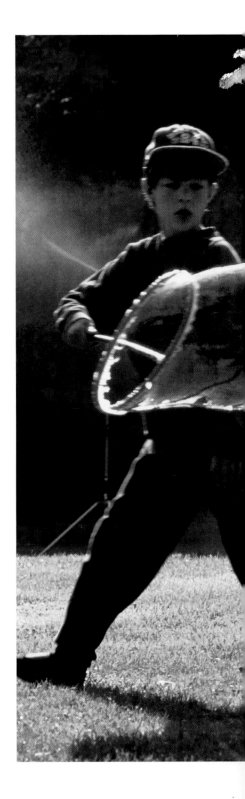

Whereas gases expand and diminish with temperature and pressure, liquids are basically incompressible. This seemingly trivial difference has profound consequences. For one thing, it makes possible powerful hydraulic systems in which small amounts of force exerted over a small area, such as the piston linked to your car's brake pedal, can produce a much larger force acting over a much larger area—the brake drums or disks of all four wheels. As a result, a gentle nudge from your foot is enough to slow and even halt a two-ton car. How is this possible? The answer lies in a law named for French scientist Blaise Pascal (1623-1662).

It states that pressure applied to a confined incompressible liquid is transferred equally to all points within the vessel containing it. That means that as long as your foot pushes on the pedal, the pressure remains the same throughout the entire system. Since pressure is force per unit of area, extending it over a larger area demands that a commensurately greater force must be applied, to keep the pressure constant.

Plants are also under a lot of pressure to perform, and they use water for structural integrity as well as nourishment. Whereas trees depend on the strength of wood to hold them up, plants lacking woody tissue rely on turgor, the water pressure within cells, to stiffen their cell walls, much as a balloon distends when air is forced inside. Cells take in water by osmosis—another aspect of Mother Nature's enduring obsession with equal distribution. ∎

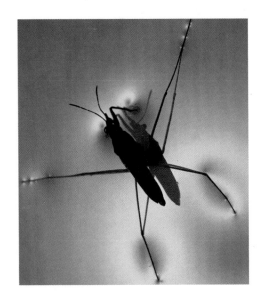

◄ Taking advantage of tension

Surface tension, the propensity of liquids to form rounded shapes, can be very supportive. As water molecules bind to one another, they act like a stretched-out sheet. Light insects avail themselves of this membrane-like quality to literally walk on water.

Bubble tech ►

Certain substances, such as soap, actually decrease the surface tension of water. In its pure form, water can't be stretched out to form a thin film. But add a little soap, and you're in business. Even then, the remaining surface tension tries to make the surface as small as possible— hence the bubble.

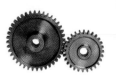

SEE ALSO

Simple tools
32-37

Adhesion
and cohesion
40-43

Pressure
and gases
44-47

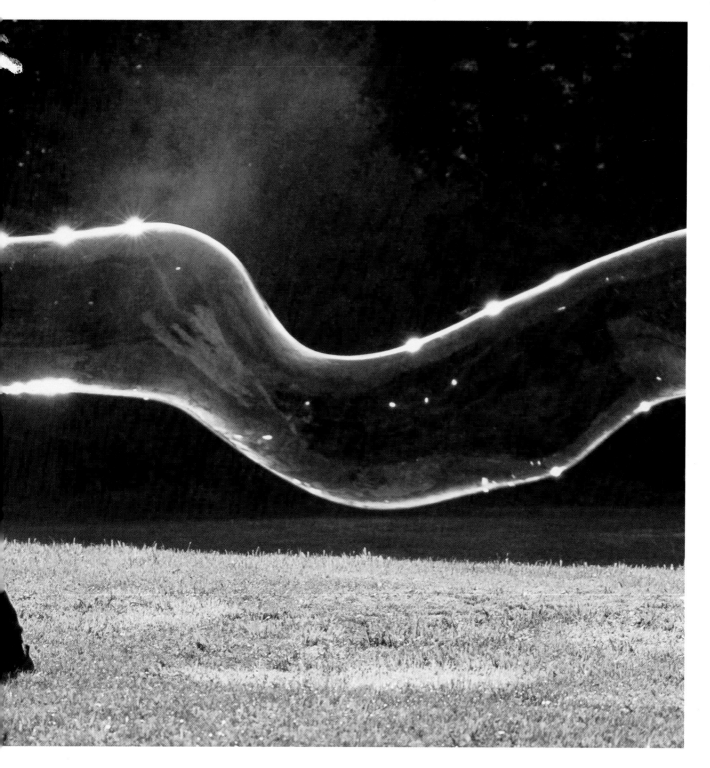

Making steel float

Intuition tells us that lighter, less dense materials float atop heavier, denser ones. But the reverse also happens, through the "magic" of displacement.

Despite the ads, there are only two surefire ways to lose weight fast: surgical procedure and immersion. The key to the latter is Archimedes' principle: An object is buoyed up by a force equal to the weight of the volume of the fluid it displaces.

Even things that sink are affected by buoyancy. That's why, should you find yourself on the bottom of a lake bed with a 150-pound rock, you'll be able to lift it with a force of only about 90 pounds. That seems odd—after all, there are several tons of water lying on top of the rock—but you can prove it just by applying Newton's laws and a little logic.

Imagine a full swimming pool, then picture a single cubic foot of the pool's water about midway to the bottom. It's just suspended there—yet gravity is pulling it downward with a force equal to its weight, which is over 60 pounds. So something must be pushing with an equal but upward force in order for the water to remain motionless. That force is buoyancy. Why does it always push up? Since pressure increases with depth, the pressure on the bottom of any submerged object—including our arbitrary cubic foot of water—is greater than the pressure at its top. The watery cube wants to rise exactly as much as it wants to fall. ■

Controlling buoyancy ▶
Once launched, submarines rise and sink according to Archimedes' principle. By varying the amount of water in their ballast tanks, they alter their overall buoyancy relative to the surrounding sea. A substance's specific gravity, or density relative to water, determines whether or not it will float. Air is less dense than water, so bubbles rise. Wood is usually less dense, so sticks and logs float.

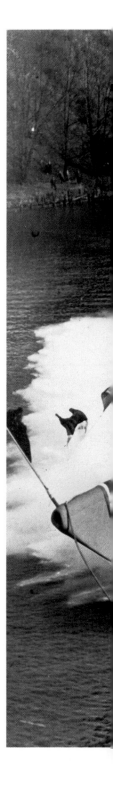

Archimedes' principle ▶
also explains why Earth's solid continents ride on seas of semimolten mantle rock below. Continental rock averages about 2.8 times the density of water; mantle rock averages around 3.3. So the continents float—although they project very little above the mantle, since the difference in densities is not extremely large.

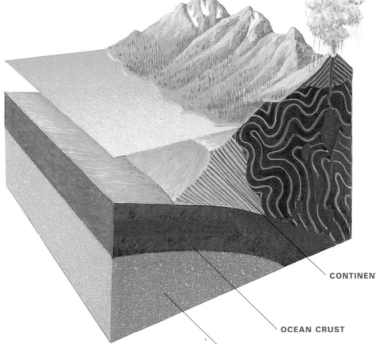

CONTINENTAL CRUST

OCEAN CRUST

MANTLE

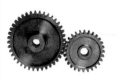

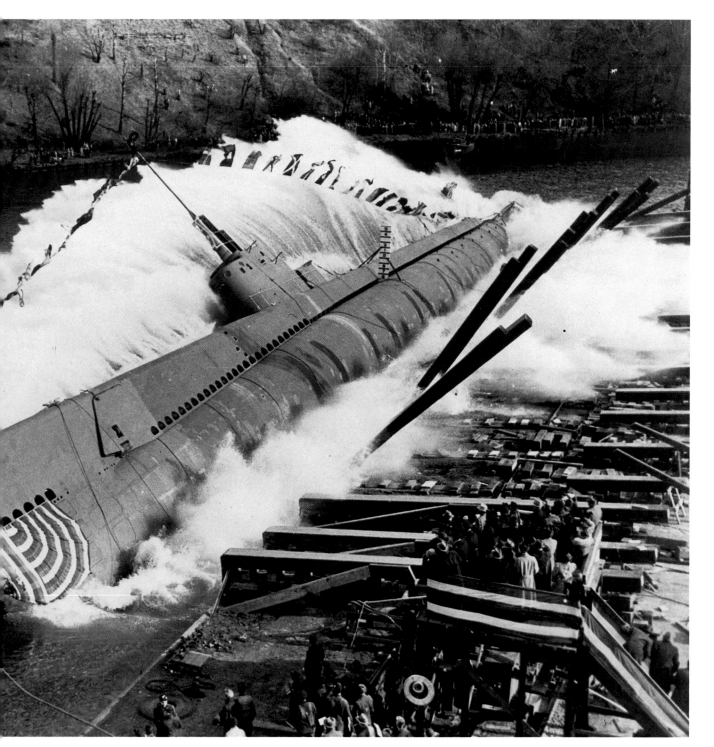

SEE ALSO

Getting started
14-15

Newton's laws
of motion
16-21

Drag, flow,
turbulence,
and lift
52-55

Gravity
and you
78-81

A turbulent world

Fluids, being indefinite of shape, give way to intruding solids—but there's a cost. Welcome to the realm of flow, turbulence, and fluid dynamics.

s any swimmer can attest, moving through fluids can be a real drag—thanks to friction-like forces that vary with the density of the medium and the shape of the object passing through it.

In fluids, smoother things often move faster. But not always: There are three basic types of fluid motion, each largely dependent on the object's speed. In the first type, called laminar flow, the fluid streamlines so smoothly around the moving object that the pressure behind it nearly equals the pressure in front. In the second type, partial turbulence, inertia prevents the fluid from following the object's contours exactly. In the third type of flow, the fluid layer immediately surrounding the moving object becomes fully turbulent.

Objects in free fall eventually reach a point at which the drag exactly offsets the acceleration of gravity. This is called terminal velocity. It's about 134 mph for skydivers, 20 mph for Ping-Pong balls, and 45 mph for basketballs.

The smoothest surface, however, doesn't always travel the fastest, especially when an object is spinning. In fact, total turbulence can have a sort of lubricating effect that helps the fluid move around the object efficiently, decreasing both the wake and the amount of drag. Though it's hard to believe, that's why sporting goods manufacturers put all those dimples on golf balls, raised stitches on baseballs, and fuzzy surfaces on tennis balls. ■

Living on the edge ▶
The world of sports is filled with racers— from formula-one drivers and motorcycle riders to downhill skiers and competition swimmers—who rely on minimizing drag through various means. Hence the clothing and helmets on these skaters. They're also "drafting"—tailgating the leader so closely that the turbulence he or she generates actually helps them to move even faster.

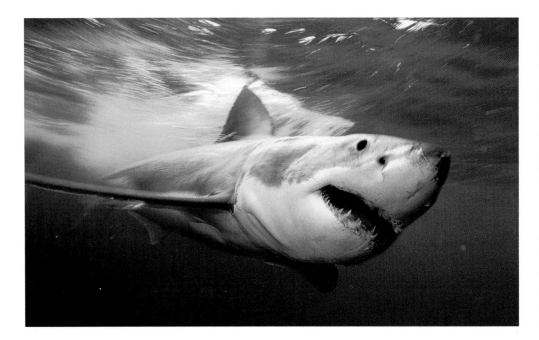

◀ Shaped for speed
Because drag varies with an object's cross-sectional area as well as its shape, it's no accident that sharks, jet planes, and late-model cars look so similar. They're all streamlined—with sleekly pointed snouts and swept-back fins—to offer minimum resistance at their normal cruising speeds in their fluid of choice.

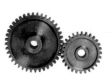

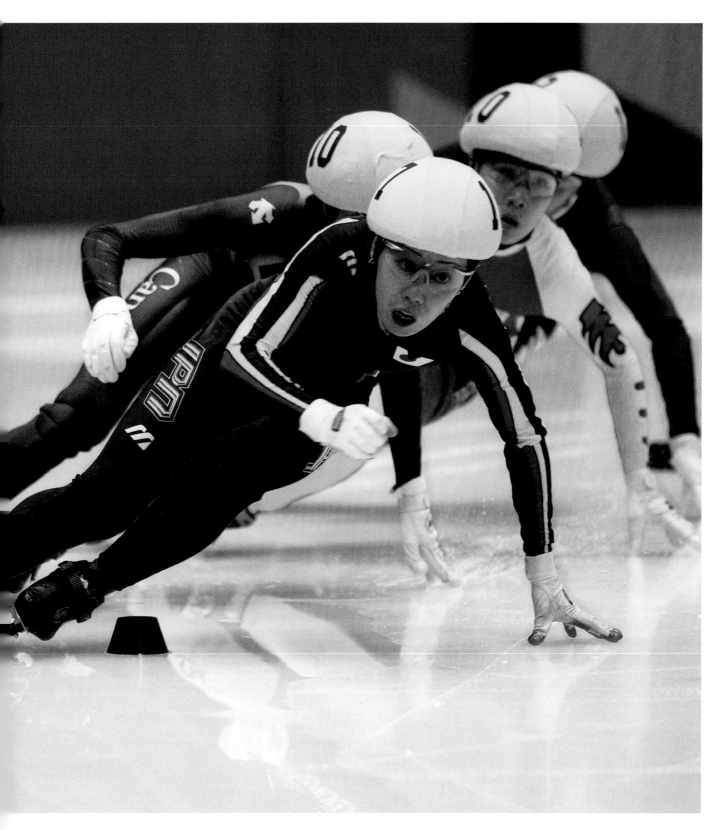

Taking Bernoulli for a ride

The principles of fluid dynamics touch us daily, whether we fly an airplane, play ball, take a shower, or strive for that extra bit of air under our skis.

First-time fliers often experience a sinking sensation when they look out the cabin window at 28,000 feet and ask themselves, "What's holding this thing up?" Fortunately, the forces that make an airplane wing rise—what we call lift—are no less powerful for being invisible. Getting a 400-ton wide-body to leave the ground or making a curveball curve both rely on differences in pressure created when objects move through a fluid—that is, through a gas or liquid. Daniel Bernoulli (1700-1782) discovered why this happens while studying the conservation of energy in liquids. He observed that water flowing through a pipe moves faster when the pipe's diameter is reduced. Since it travels faster, some force must be acting on it. That force, Bernoulli reasoned, must arise from differences in pressure: The slower fluid in the wider part of the pipe must have higher pressure than the faster-moving fluid in the narrow part. That would mean that the pressure of a liquid or gas is lowest when its velocity is highest and vice versa. Application of such principles gives lift to aircraft (as well as to birds and to ski jumpers). It also explains why bits of paper or cigarette smoke are sucked out the open window of a fast-moving car, and why your shower curtain billows inward, toward the faster-moving air pushed by the shower spray, rather than outward, when you turn on the water. ∎

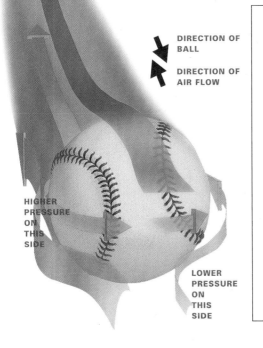

DIRECTION OF BALL

DIRECTION OF AIR FLOW

HIGHER PRESSURE ON THIS SIDE

LOWER PRESSURE ON THIS SIDE

Giving a baseball that certain spin

When a pitcher throws a 70-mph curve, the ball rotates about 17 times in its 60-foot journey to home plate. As it revolves, its 216 raised stitches drag a thin layer of air around the ball. If the ball spins counter-clockwise (as seen from above), the side toward third base travels against the main flow of air, while the side toward first base moves with it. Thus the speed of the air relative to the ball's center is different on each side. That causes the air pressure to be relatively lower on the first-base side and relatively higher on the third-base side. As a result, the ball curves toward first—a striking demonstration of fluid dynamics.

Getting that essential lift ▶
Birds and planes fly for the same reason: Their wings form airfoils that cause the air flowing along their top surfaces to move faster than the air below, making the pressure lower above the wing than under it. That's lift. Pilots can vary the degree of lift versus drag by extending or retracting flaps on the wings' trailing edges; birds accomplish the same thing by adjusting their long primary feathers.

Preventing takeoff ▶
At high speeds, streamlined racecars begin to rise. That's just what drivers don't need, since they depend on friction between tires and asphalt to run safely. So they (and many hot sports cars) are designed with "spoilers"—basically, upside-down wings—mounted front and rear. As a result, air pressure "lifts" the cars downward, pushing them ever harder into the track the faster they go.

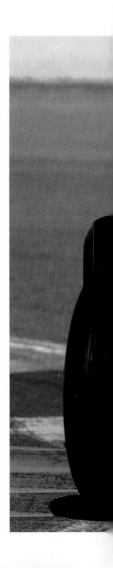

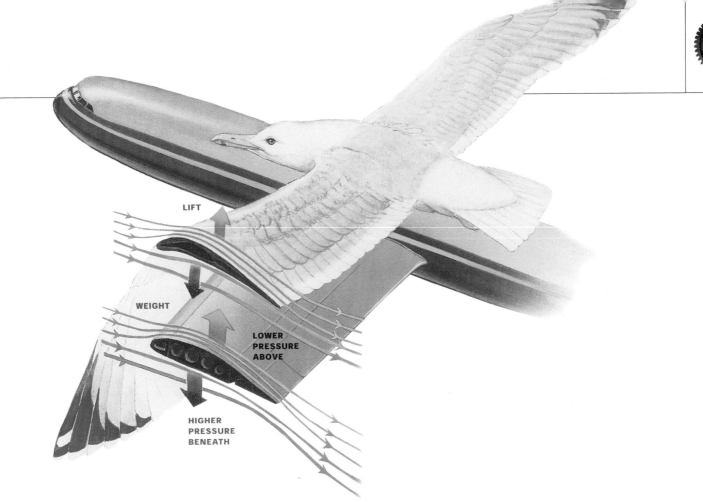

LIFT

WEIGHT

LOWER
PRESSURE
ABOVE

HIGHER
PRESSURE
BENEATH

SEE ALSO

Friction
as friend
28-29

Pressure
and gases
44-47

A turbulent
world
52-53

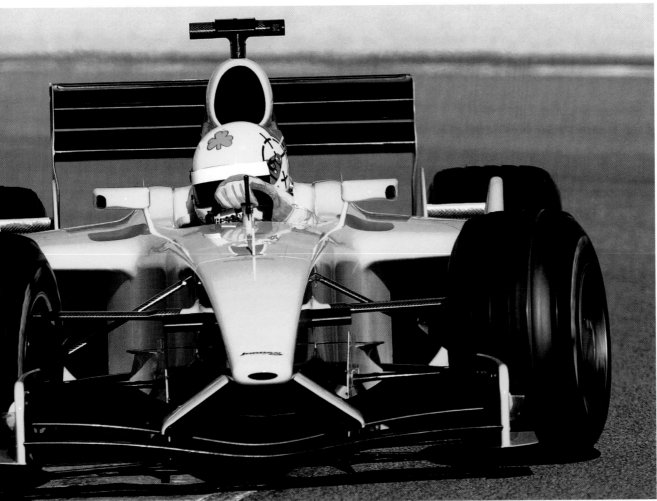

Making waves

Surprising as it seems, sometimes the thing that's moving isn't a thing at all. It's a wave: a traveling disturbance or fluctuation.

We've all seen it: two children holding a jump rope. Suddenly, one shakes his end, causing a kink or loop that speeds over to the other child. But what, exactly, was transferred from one to the other? Obviously, not the rope itself, since both it and the kids remain right where they were. Only the kink—a temporary disturbance in the rope's state—actually changed position. Humans have spent centuries trying to comprehend such moving undulations in the cosmos, which range from the fury of storm-whipped surf to the alternating compression and thinning of the air that is the sound of a cello, to the disembodied rays that carry sunlight, as well as radio, television, and other signals. We call all these things waves.

Although they often appear to have little in common physically, all waves have certain essential features. They are cyclical, like ocean waves. The total distance covered by one complete cycle—from peak to peak or trough to trough—we call wavelength; the number of waves that pass a stationary point per unit of time is termed frequency. ∎

It's all about energy

All waves convey energy. Sometimes the amount of energy in a wave depends on how hard it compresses the medium through which it travels—as occurs with sound, ocean, and seismic waves. But then there's light, which is made up of individual wave forms, or energy packets, called photons that can pass through a vacuum as well as a medium such as air, glass, water, or anything else that's transparent. To pass through more dense materials, such as flesh and bone, it takes light waves of smaller size and higher frequency than visible light. X-rays can do the job.

What is a wave? ▶
The strength or size of each wave is its amplitude. Its speed equals the product of frequency times wavelength. In light waves, the higher the frequency, the higher the energy.

The tsunami punch ▶
So-called tidal waves, or tsunamis, have nothing to do with tides. They stem from seismic events, such as earthquakes, and consist of high-speed sea waves that have crests of enormous height.

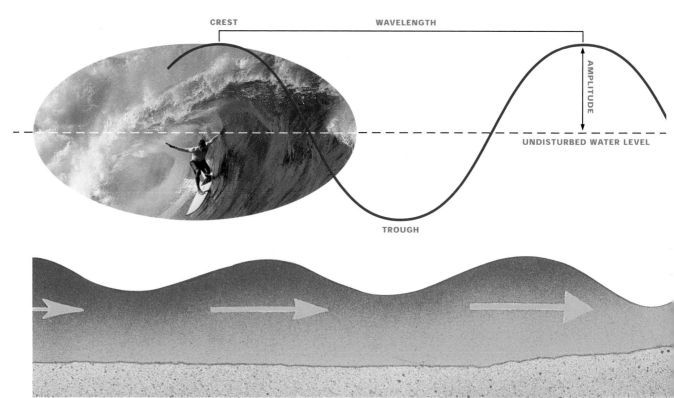

CREST   WAVELENGTH

AMPLITUDE

UNDISTURBED WATER LEVEL

TROUGH

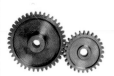

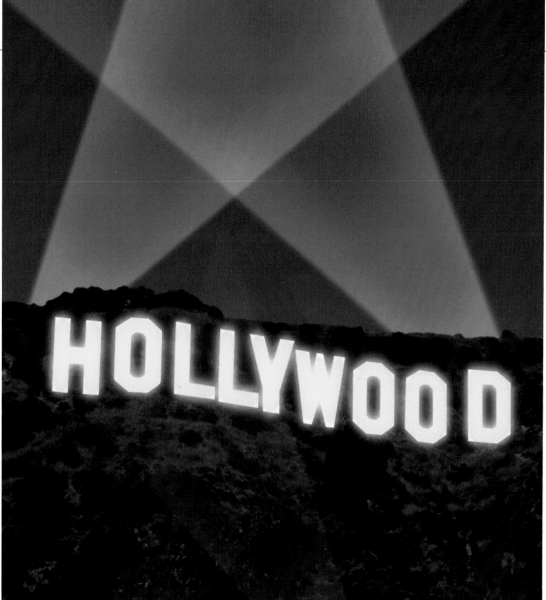

Show time ▶

In life as in nature, waves make any medium more exciting. The energy of light ranges over a huge scale, from low-frequency radio waves that pass harmlessly through your body every second to ultra-powerful gamma rays produced by the most violent events in the cosmos. Visible and heat radiation from klieg lights merely warms your skin; an equivalent dose of photons from high-frequency gamma rays would fry you.

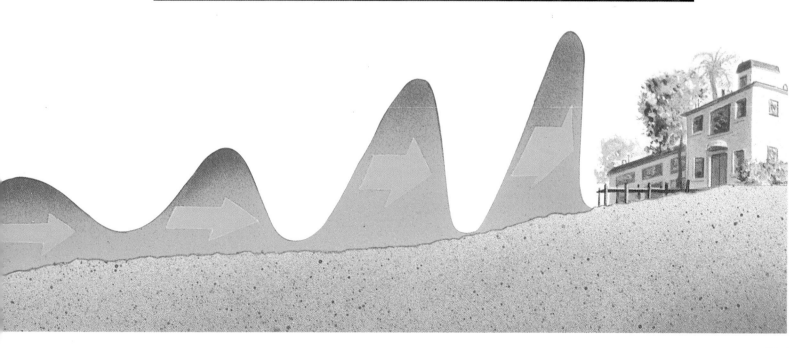

Blown away

This advertisement for a manufacturer of audiotapes whimsically dramatizes the power of sound. Unlike ocean waves, which move up and down, acoustic waves are disturbances in density, alternately compressing and then rarefying the medium through which they travel. Strike the surface of a bass drum, and it vibrates in and out. Each outward bulge compresses the air immediately next to it, each inward motion causes a rarefaction. Sequential chains of such high- and low-density air masses comprise the sounds we hear, which range from a few dozen cycles per second to about 20,000 cycles per second. Like all waves, sound moves at speeds that change with the density of the medium. While it averages about 740 mph through air, it can race through steel at speeds of more than 10,000 mph.

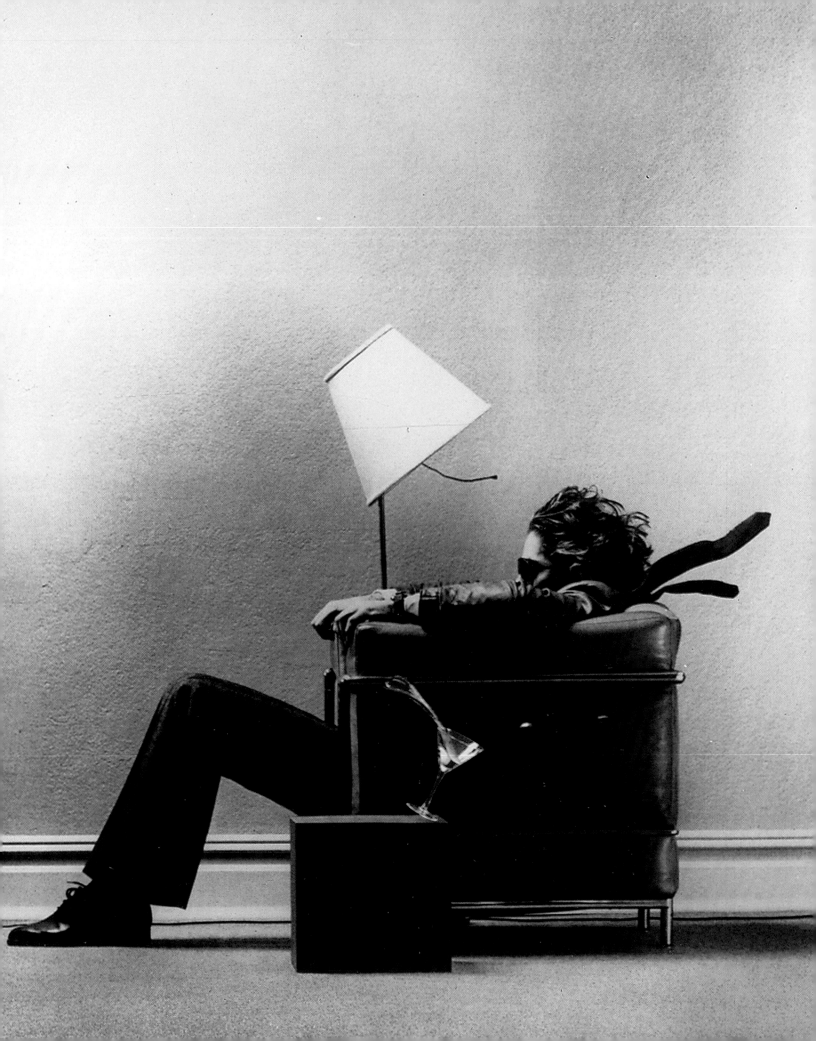

Reflection and refraction

In our wave-swept world, we have learned to exploit all sorts of waves for our everyday benefit. Bouncing and bending them are key methods.

Dozens of modern devices rely on the simple fact that whenever waves pass from one medium to another, they either bounce back or bend—that is, reflect or refract—in predictable ways that can be utilized to create images of things that otherwise would remain invisible to us. For example, we know that the time it takes a sound to reflect off an object and travel back to its source is proportional to the total distance involved. Submarines employ this principle every time they use sonar to gauge their distance from underwater formations or other ships. That's the same basic method we use to estimate our distance from a lightning bolt: Tick off the seconds between the strike and its thunder, and divide by five to get the approximate mileage. Since sound travels about 1,100 feet per second in air, that's just over a mile every five seconds. (Sound moves four times faster in water, thanks to the liquid's higher density.) Weather radar sends out electromagnetic waves of about the same frequency as your microwave oven, with a wavelength of a few inches. The reflected signal, captured by the receiver, shows the shape of the storm. ■

▼ The nature of quakes

Worldwide networks of seismographs enable scientists to track natural earthquakes, pinpointing places of origin and determining relative strengths. Quakes promulgate two kinds of motion at once—compressional "primary" or P waves, and up-and-down "secondary" or S waves.

In general, S waves cause more damage to man-made structures, because they shake foundations. Many earthquakes follow natural faults between two geological formations. When faults are heavily stressed—as often occurs wherever Earth's massive tectonic plates collide or grind past each other—any slippage or fracturing along plate edges may release the trapped energy, triggering a quake.

Wave patterns ▶

As waves travel through a medium, they can lose a lot of energy. Look at the difference in amplitude (water height) of this wave pattern as it moves from the center of impact outward, becoming weaker as it gets farther away.

SECONDARY WAVES

PRIMARY WAVES

SEISMOGRAPH READING

YELLOW ARROWS: DIRECTION OF WAVE MOVEMENT

RED ARROWS: DIRECTION OF ROCK MOVEMENT

FAULT LINE

FOCUS

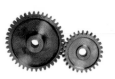

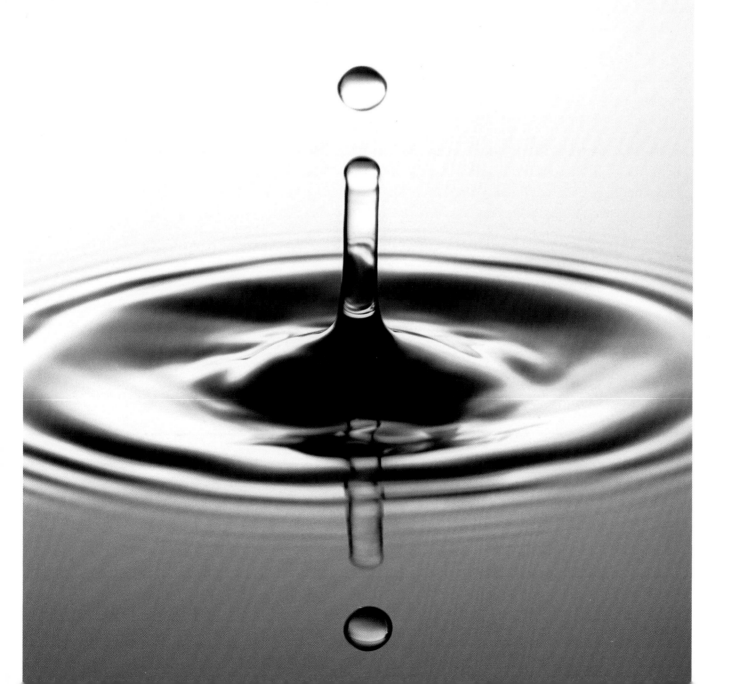

Resonance and interference

Like people, waves interact both in constructive and destructive ways. Depending on what you want to achieve, either path can be an asset.

When two waves of the same length and amplitude meet, they reinforce each other as long as they're in precisely the same stage of their cycle—that is, if they perfectly overlap. Such waves are said to be "in phase." They demonstrate constructive interference, for together they act as a single wave whose amplitude is the sum of the two component waves.

But if these waves meet in such a way that the trough of one aligns exactly with the crest of the other, the two will totally cancel each other. That's destructive interference. But just because it's called "destructive" doesn't mean it's all bad. Some entrepreneurs, for example, have used the principle of destructive interference to market devices that fight noise with noise. These machines detect irritating sounds in a room, such as an air conditioner's hum, then generate sounds identical in frequency and amplitude but exactly out of phase. The two cancel each other out and you hear nothing.

When two groups of waves are only partly in phase or of different wavelengths, combining them results in a pattern that sums the characteristics of both waves at any given point. Radio stations do this when they combine the wave patterns from music or talk with their particular "carrier" waves. ■

▼ **We've all heard this one**
Ever notice how the pitch of an ambulance or police siren rises as it approaches and drops as it passes? That's because when the siren moves toward you, the vehicle's motion pushes the waves a bit closer together, raising their frequency. But after it has roared by, the vehicle's velocity puts slightly more distance between each wave, producing sound of a lower frequency. This phenomenon is called the Doppler effect. Meteorologists use it to see how different parts of a storm are moving.

In the wake ▶
This image nicely captures the combination of constructive and destructive interference. As the wave pattern of the boat wake combines with the larger pattern of ripples on the water surface, it creates alternating areas of high and low amplitude.

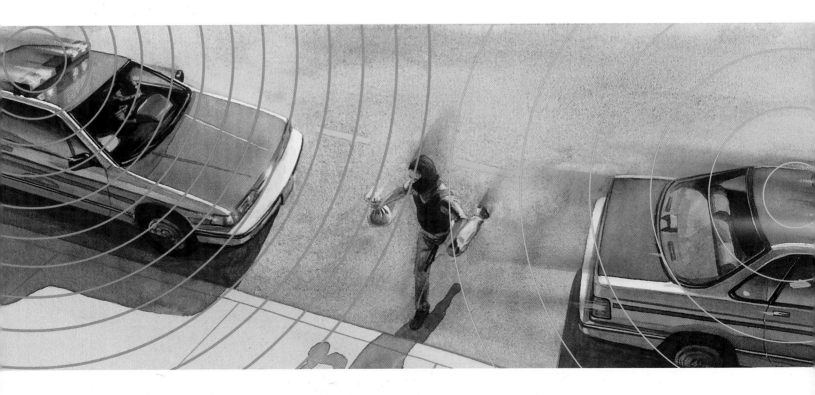

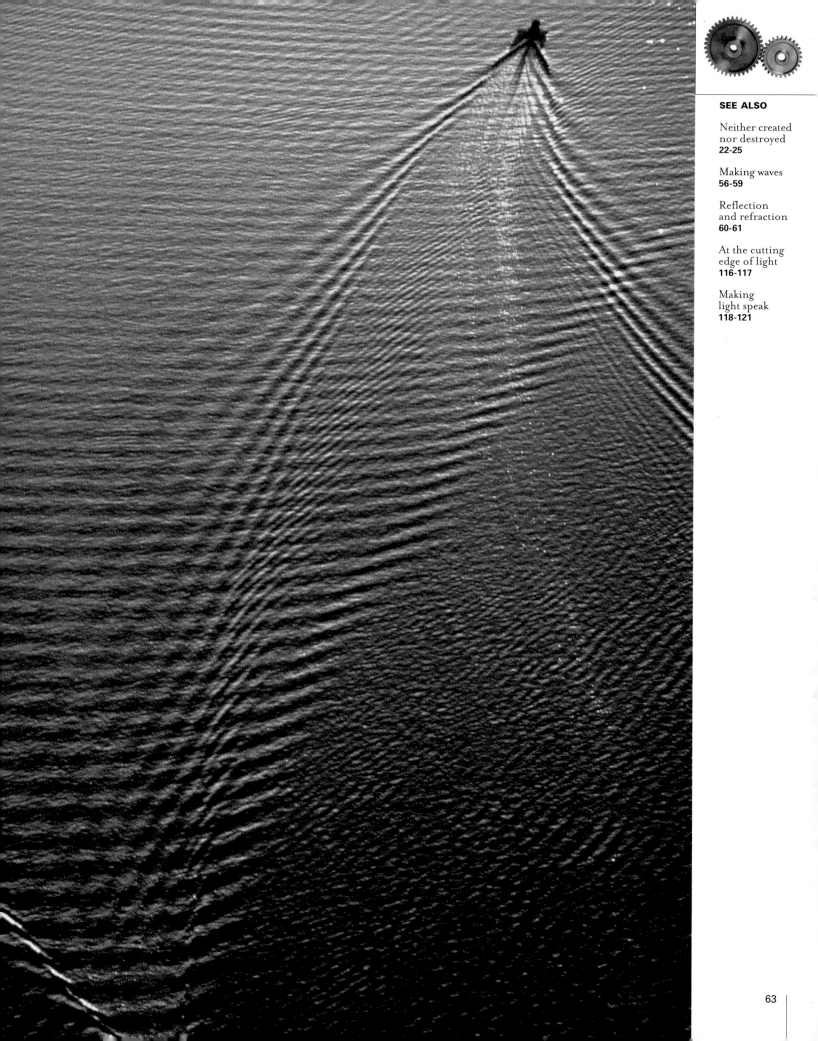

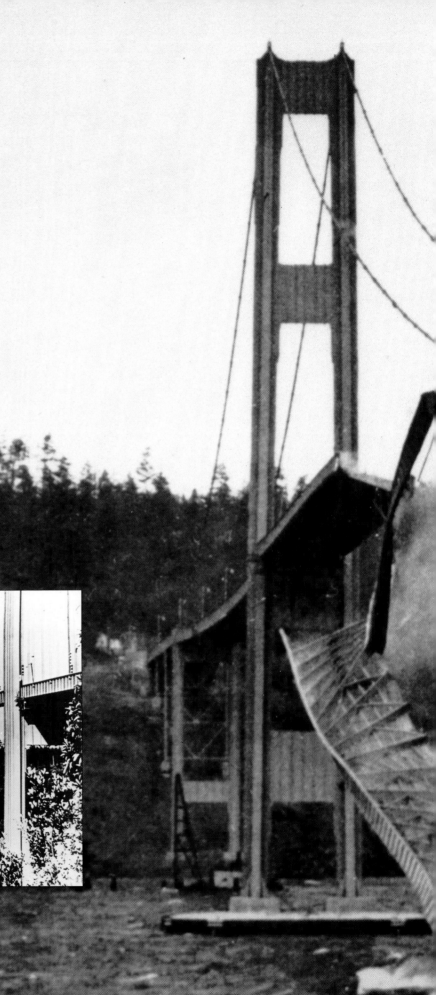

Before the fall

Resonance doesn't occur only with sound waves: Winds also can precipitate it. That's what befell Washington State's Tacoma Narrows Bridge in 1940, after gusts at just the right frequency caused the single-suspension span to twist (inset), then literally shake itself apart. Such hazards of resonance are well known to the military; some commanders routinely march their troops out of step across bridges, as synchronized footsteps could match a structure's resonant frequency and trigger a collapse.

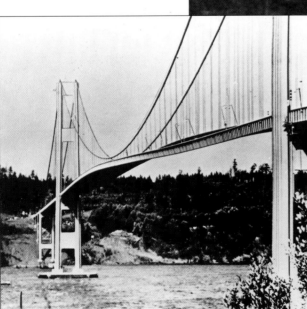

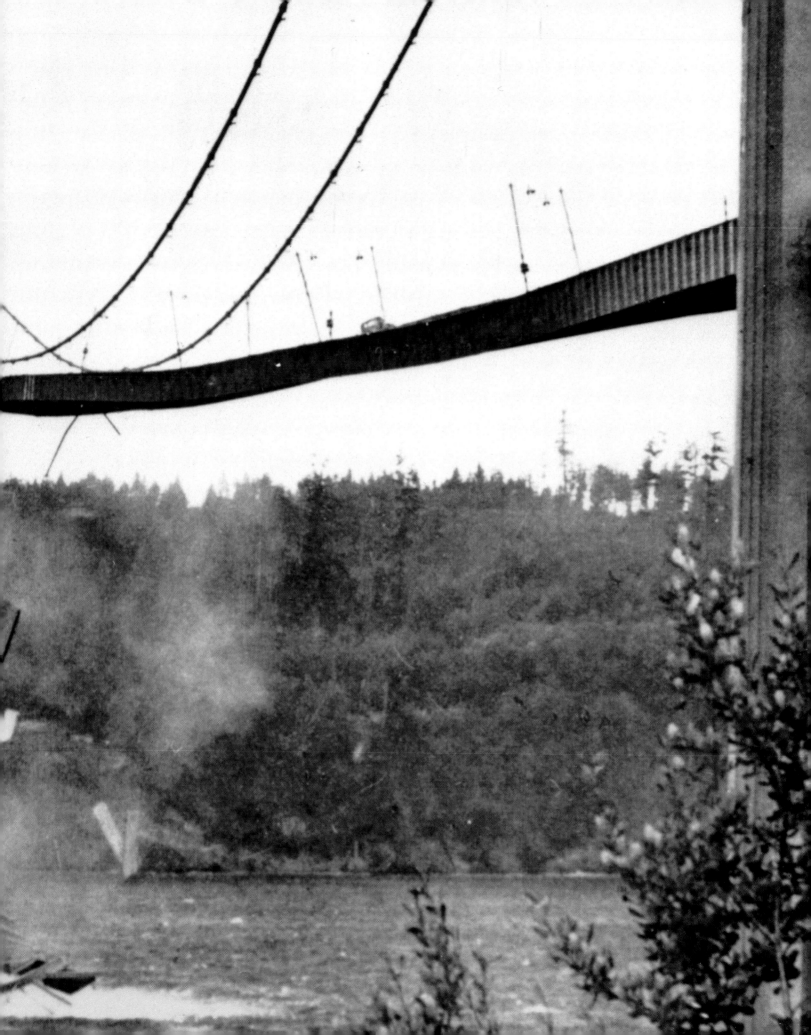

Turning up the heat

If waves are nature's paramount expression of rhythmic motion, then heat is the very essence of disorganized, random energy.

Heat is raw energy, and it boils down to matter in motion. Atoms and molecules, the building blocks of everything around us including ourselves, move constantly and randomly; the faster they move, the warmer the substance they make up. Every object in this world—no matter how frigid it may seem—contains some heat. Even a jug of ice water harbors so much molecular motion that if you gently place a drop of ink on the surface, it will diffuse evenly throughout the liquid. In theory, there's a point called absolute zero where all motion—and hence all heat—ceases to exist. But it remains unattainably cold: -460°F.

Heat always travels in whatever direction tends to equalize temperatures; that is, from regions of high thermal energy and relative warmth to colder areas. Bring enough heat together in one place, and you may be able to overcome the forces of attraction between atoms and molecules, causing a change of state from solid to liquid or liquid to gas. Such changes require additional energy, called latent heat, which doesn't raise the substance's temperature but is needed just to change state. Consequently, it takes more energy to raise the temperature of a quart of water from 210°F (liquid) to 220°F (steam) than it does to raise its temperature from 85°F to 95°F. ■

Starring plasma ▶
As they gain heat, and their constituents move farther apart, substances go from solid to liquid to gas. Beyond that, there's another state of matter—called a plasma—in which there's so much energy that electrons can't even stay attached to their atoms. The resulting miasma of charged particles is the normal condition of gases in stars.

◀ Microwave magic
Cooking without a flame, microwave ovens work by producing waves that are just the right frequency to cause water's highly polarized molecules to flip around faster and faster (inset). Since even "dry" foods often contain moisture in the form of water vapor or chemically bound water, the oven warms them as well— and manages to boil coffee without heating the cup.

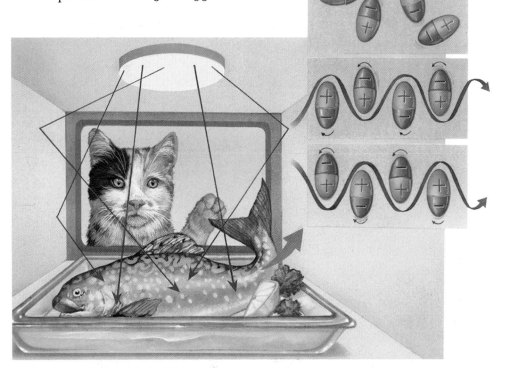

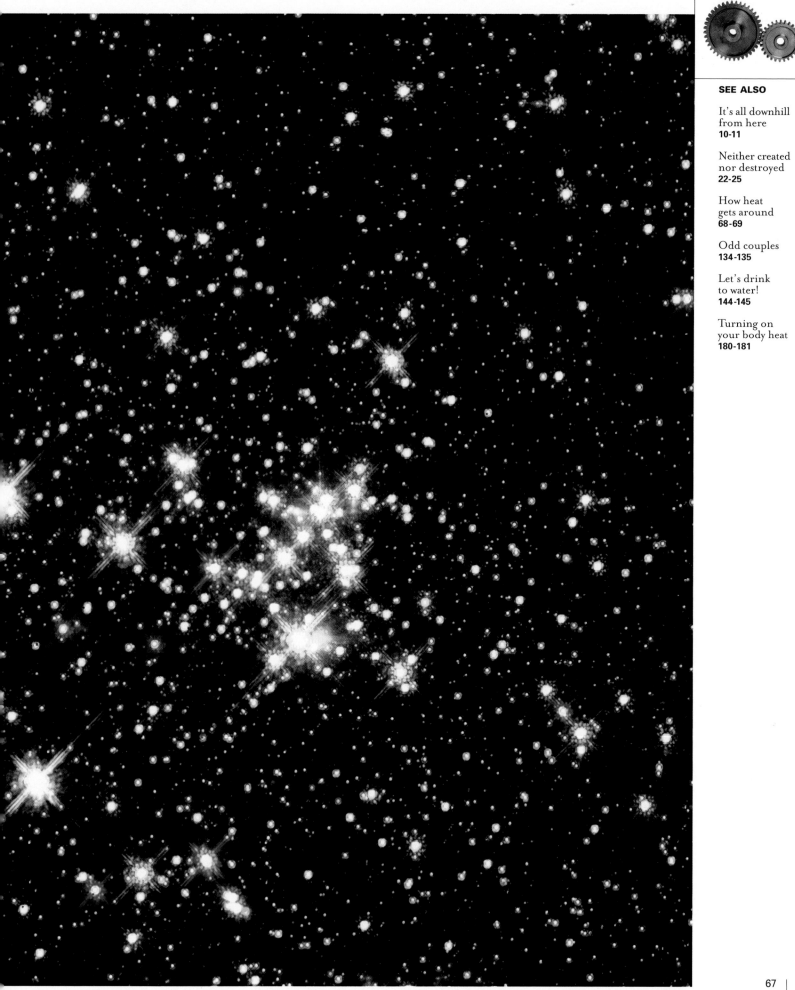

SEE ALSO

It's all downhill
from here
10-11

Neither created
nor destroyed
22-25

How heat
gets around
68-69

Odd couples
134-135

Let's drink
to water!
144-145

Turning on
your body heat
180-181

How heat gets around

Heat, being not matter but raw energy, can change into other forms. On its own, it travels by three basic methods: conduction, convection, and radiation.

Throughout most of the world, the basic unit of heat is the calorie, which is the amount of heat necessary to raise the temperature of one gram of water one degree Celsius. We also use the word "calorie" to describe the energy content of food. But food calories—often spelled with a capital C—are actually kilocalories, a unit one thousand times larger.

Any given quantity of heat can have very different effects on different materials. For example, it takes five times as many calories to raise the temperature of water by one degree as it does to warm an equal mass of marble, aluminum, or glass by the same amount. This property is called specific heat, and you feel its effects whenever you visit the beach. When the sun shines, land heats up faster than ocean, so the air over land gets hotter and rises. Air above the comparatively cooler ocean rushes in to fill the space, which explains why sea breezes generally blow onshore during the day. At night, exactly the opposite transpires: Land cools faster, making the air over the ocean relatively warmer and prone to rise. The breeze then blows from shore to sea.

Unlike other methods of heat transfer, radiation doesn't require a medium. Heat energy in the form of infrared electromagnetic waves travels quite nicely through nothing at all, which is how heat gets to Earth from the sun. Or to you from your hot-water radiator, glowing space heater, or fireplace hearth at home. ■

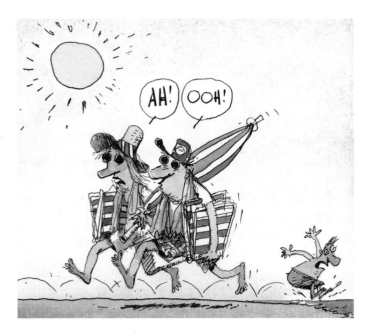

Foiling heat loss

Insulating materials slow down the process of thermal energy transfer. The walls of your home may contain blankets of fiberglass with a shiny foil layer pointed toward the living space. The foil's job is to reflect radiated heat back into the rooms rather than let it dissipate through exterior walls. Beneath the foil, fibers create a layer of trapped air—a notoriously poor conductor of heat—thus retarding any heat loss to the outside by conduction. The styrofoam in your picnic cooler works just the same way.

Convection ▶

The density of a fluid decreases as it is heated. Its constituent molecules spread farther and farther apart, making the fluid more buoyant. Whether you use radiators, air ducts, or baseboard heaters, your home spreads heat by convection: Warm air rises to the ceiling, gradually loses some of its heat to its surroundings, grows more dense, then descends.

◀ Conduction

When atoms and molecules pass thermal energy by direct contact, that's conduction. It occurs more readily in solids than in liquids or gases because the molecular structure is denser. Atoms of the best conductors are joined in orderly arrays that enhance energy's ability to travel. Metals are particularly good: Silver and copper conduct heat about nine times better than steel; and steel, is 50 times better than glass or brick.

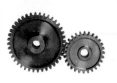

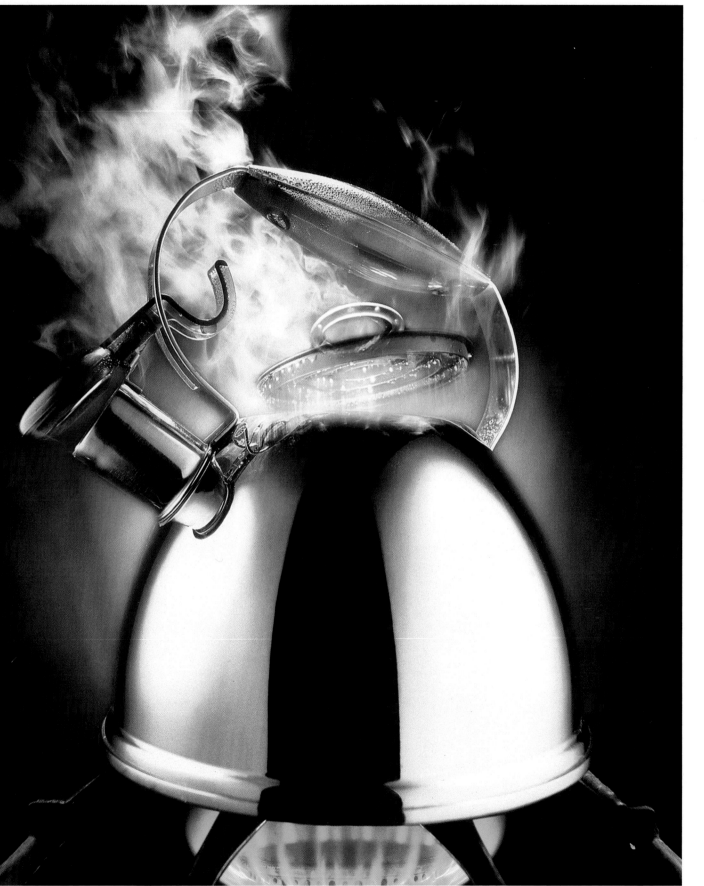

Heat and weather

Any day of the year, atmospheric forces of convection can create thunder-heads tens of thousands of feet high and hurricanes hundreds of miles wide.

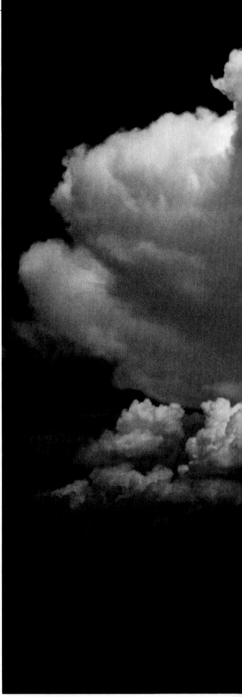

A major key to the weather is convection, powered by the impact of the sun's radiation on Earth. On an average day, every square yard of our upper atmosphere directly facing the sun receives an amount of radiation equal to the output of about ten 100-watt lightbulbs. But only a fraction of that total ever reaches the planet's surface.

How much winds up warming the lowest level of the air? That depends on several variables, such as whether the radiation falls on water—with its high capacity for holding heat—or on land, with its comparatively low capacity. Cloud cover also is important, since clouds block part of the sun's energy. The meteorological bottom line is this: Some areas will always have hotter surface air than others. Intense surface heating causes the rising air column to punch up farther and farther into the atmosphere, which is why thunderstorms occur most frequently in the tropics.

The simplest way to think about the behavior of Earth's fluids—air and water on the surface, molten magma deep in the interior—is to regard the entire system as a gigantic mechanism that seeks to achieve energy equilibrium by moving heat from areas of high concentration to low. But the surface can never reach equilibrium because the sun heats only half the planet at a time and warms each hemisphere differently depending on season. ■

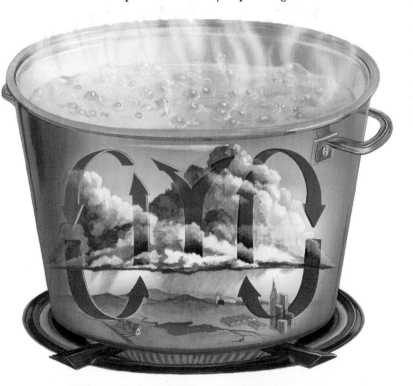

Boil and bubble ▶

Like water in a gigantic cauldron, Earth's atmosphere is in constant motion. Warm air, with lower density and greater buoyancy, rises—but only until its density equals that of the surroundings. When it cools, it loses its capacity to hold water vapor—which begins condensing into clouds. The cooled air descends—until it is warmed again.

▲ Nature's fireworks

Convection builds thunderheads that are five or more miles high. As air, water drops, and bits of hail swirl in the cloud, collisions of particles cause electrical charges to get separated—with positive at the top and bottom and negative in the middle. When the difference reaches 100 million volts or so, it overcomes the normal electrical resistance of air, and lightning strikes.

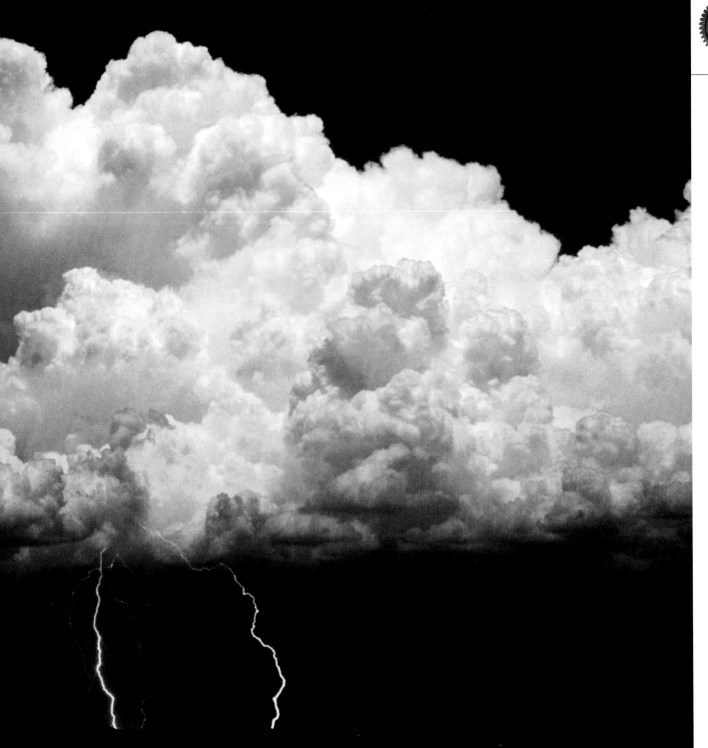

Temperature inversions

The troposphere—the air in the midsection of Earth's atmosphere—normally drops in temperature about 28°F for every mile of altitude. But sometimes, the weather drags in a warm layer that sits atop surface air like the lid on a pot. Convection is thwarted, and civilization is obliged to stew in its own pollutants until something happens to dislodge the lid. That's the meteorology behind a temperature inversion.

The effects of Coriolis

Just because we're all passengers on spaceship Earth doesn't mean we travel at the same speed or even in the same class. Our vehicle, after all, is a sphere.

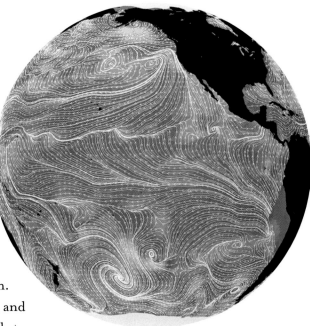

Earth takes us for a quick spin every day, but just how quick depends entirely on where we live. The Equator rotates at nearly a thousand miles per hour, but most parts of the United States move just over half that fast. The poles are all but motionless with respect to Earth's axis. Such tremendous differences lie at the heart of the Coriolis effect, named for French physicist Gaspard Coriolis (1792-1843).

When a mass of equatorial air, moving from west to east, veers north to fill a low-pressure pocket, its momentum drives it east of due north. Conversely, a mid-Atlantic air mass heading south will lag to westward of its faster-moving destination. Either way, both air masses veer to the right in the Northern Hemisphere, and the left in the Southern.

So when hot oceanic air rises in the Atlantic north of the Equator, and surrounding air on all sides rushes in, each incoming bit bends slightly to the right, forming a counterclockwise pattern that is the distinctive shape of a hurricane or tornado. ■

Down the drain ▶
Contrary to persistent popular mythology, water draining from a sink or bathtub does not rotate in different directions in the Northern and Southern Hemispheres. At kitchen and bath-size scales, any possible faint Coriolis effect is far exceeded by small irregularities in the shape of tub or sink, and by motion patterns in the water as it gurgles away.

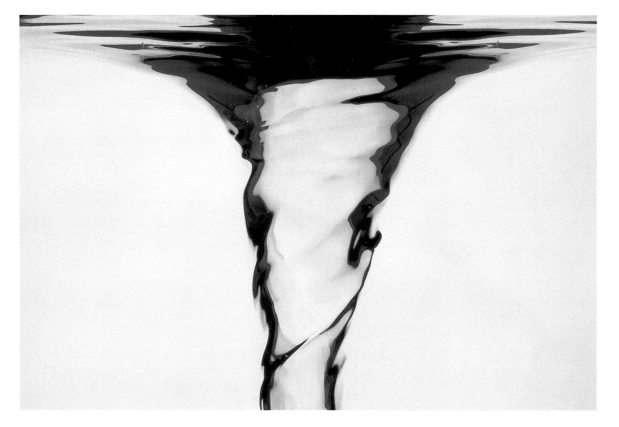

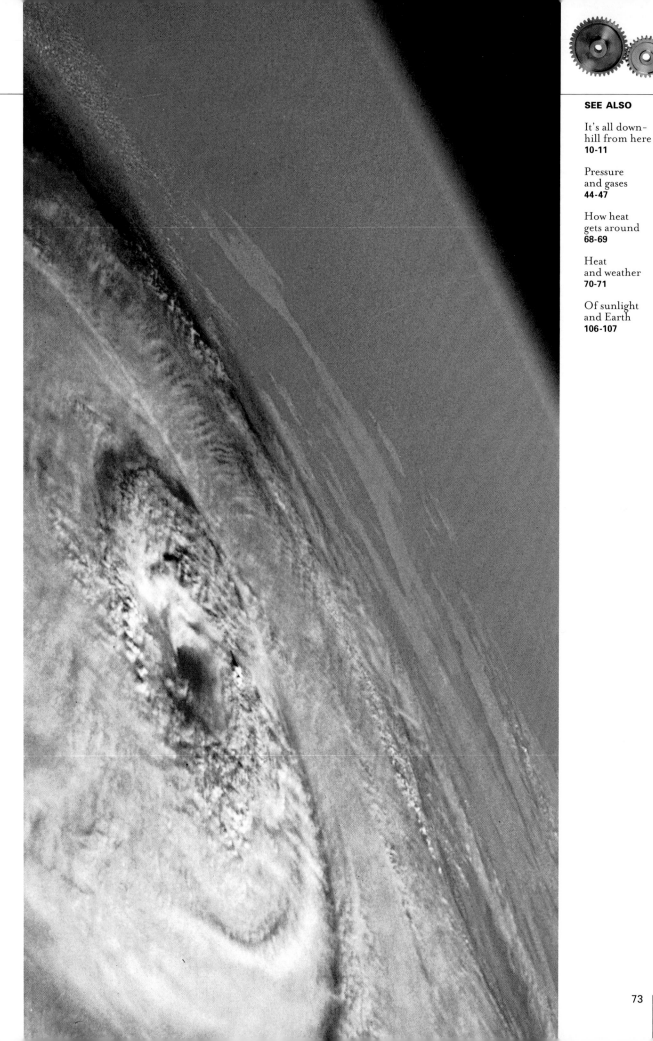

◄ Separate hemispheres

As warm, moist air near the Equator rises, it cools; pressure builds aloft, exceeding the pressure of similar air masses to the north or south. That's why equatorial air generally moves toward the poles; very little mixing occurs across the Equator, as this satellite image shows. In fact, weather patterns within each hemisphere normally fall into four latitudinal regions.

Eye of the storm ►

Northern Hemisphere hurricanes often form in the Atlantic Ocean between 10 and 30 degrees north of the Equator. Plentiful tropic heat from solar radiation raises the temperature and lowers the density of air near the ocean surface, causing it to rise. As cooler air rushes in beneath the rising mass, it begins to rotate counterclockwise.

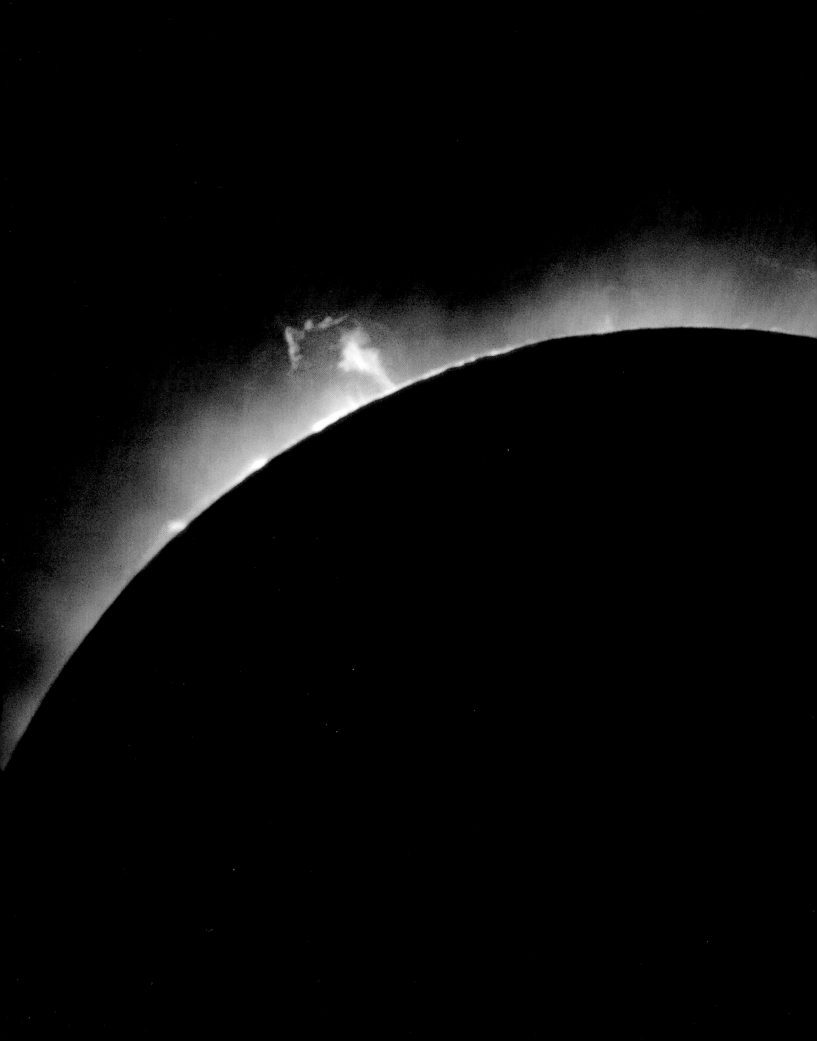

Forces
of
Nature

Contemporary physicists have reduced the awesome intensity and complexity of the entire cosmos to a mere handful of building materials—called elementary particles—and to four fundamental forces that hold everything together. The tempestuous blazing of the sun, the delicate glimmer of a rainbow, the doomsday blast of a lightning bolt, the final plunge of a great waterfall—all, as incredible as it may seem, are mere variations on those few basic themes. Yet to some, even four very different forces seem too many; Einstein spent most of his life trying to interpret all known forces as different aspects of a single superforce. The quest for a grand unified field theory continues even now.

The four forces

Just as the apocalypse has its four horsemen, nature has the interplay of its four basic forces. Science is trying to consolidate them even more.

Of nature's four fundamental forces, only two are immediately obvious in daily life: gravity and electromagnetism. The other two are just as essential—without them, we'd collapse into puddles of particles—but they operate exclusively on subatomic levels. These are the so-called strong and weak forces, responsible for nuclear fusion and fission, among other things.

Gravity is the obvious but enigmatic force that pulls everything together. Acting over tremendous distances, it keeps galaxies from whirling apart and binds the universe into the dependable patterns that make up the night sky.

Electromagnetism unites two entities once thought to be separate. Today, we see electricity and magnetism as two halves of a single electromagnetic force. It was not until the 19th century that scientists found that electrical charges in motion produced magnetic fields, and that varying a magnetic field would, in turn, cause an electrical current to flow. We have since capitalized on this eternal shoving match by creating motors, generators, computers, cell phones, and other newfound "essentials" of human existence.

The four fundamental forces are about as close as nature gets to pure magic. Each produces a disembodied force field—a term straight from science fiction but actually commonplace reality—that affects objects without exactly touching them. Each has its own carrier. For electromagnetism, that's the photon: a massless entity that is both particle and wave. Gravity is thought to be embodied in a mysterious unit called the graviton. And science is incessantly searching for more particles and fields. Stay tuned .... ∎

▼ That sinking feeling
Stand on a scale the next time you take a high-speed elevator, and you'll see dramatic weight swings. Punch the "up" button, and the car's acceleration—coupled with gravity's downward pull on your body—makes you feel much heavier. Press "down," and your head aims for the ceiling. Both situations illustrate the relationship between acceleration and weight.

Fields of force ▶
We only see the "shape" of invisible force fields when they force visible matter into a pattern. Here tiny iron filings help reveal the shape of the magnetic field lines surrounding a bar magnet. The field is always strongest where the field lines are closest together—in this case, at the ends of the magnet. Electrically charged particles tend to move along field lines, which is what produces the giant loops of plasma that bulge off the sun like incandescent ropes.

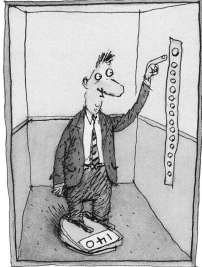
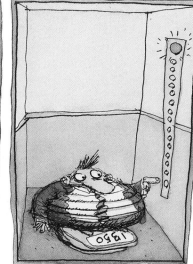
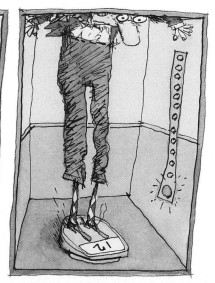

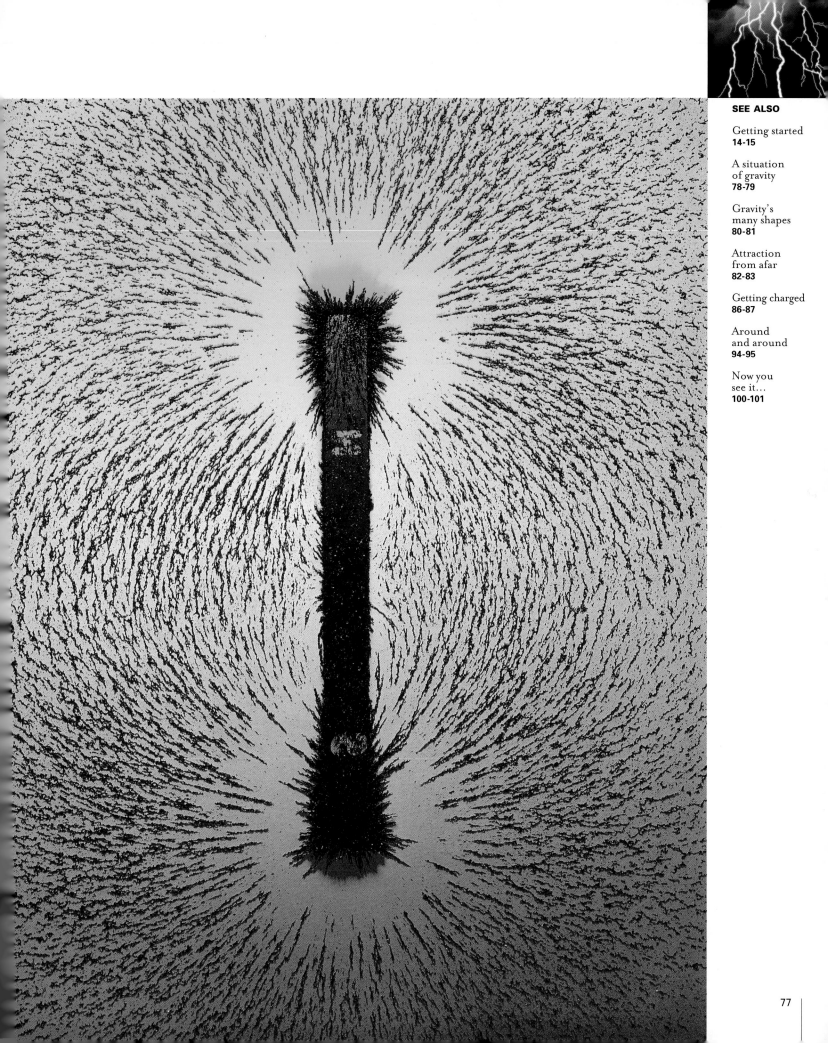

A situation of gravity

Like a big-city mayor, gravity has a lot of pull on the local level. Its influence diminishes rapidly with distance, however—though it never drops to zero.

ravitational force is proportional to the masses of the objects involved, but it decreases with the square of the distance between them. So if you gain mass in the form of, say, cheesecake, Earth's gravitational field pulls harder on the new, bulkier you—and your gravitational field pulls harder on Earth—a change that registers as increased weight on your bathroom scale. Of course, if you could move twice your present distance from the planet's center of mass, the mutual attraction would only be one-fourth as great.

Gravity's interrelationship with acceleration underlies the g-force, a concept that affects jet pilots and roller coaster riders alike. From Newton's second law of motion, we know that when an object is rapidly accelerated, some correspondingly large force must be at play. If you're on an airplane that suddenly falls and then pulls back up again, you experience a momentary sensation of weighing much more than you normally do. That's because the upward accelerating force needed to pull out of the fall is stronger than gravity alone. One way of expressing the plane's accelerative force is in "g" units, since gravity causes falling objects to accelerate at 32 feet per second per second near the Earth's surface. A baseball pitcher imparts an acceleration of about eight g's to a 100-mph fastball.

Astronauts in orbit are not actually "weightless." They are constantly pulled toward Earth with a gravitational force that just balances their momentum. But so are their surroundings, so they seem to "float" in the space station. ■

In for a landing ▶
Gravity tugs a high-flying long jumper back to Earth at the end of her leap.

▼ Prince of tides
The moon's gravitational field causes the tides. Tides are highest during new and full moons, because in those cases the moon aligns directly with gravitational pull of the sun. Conversely, tides are lowest when the moon is half full, because its gravitational attraction acts at right angles to the sun's.

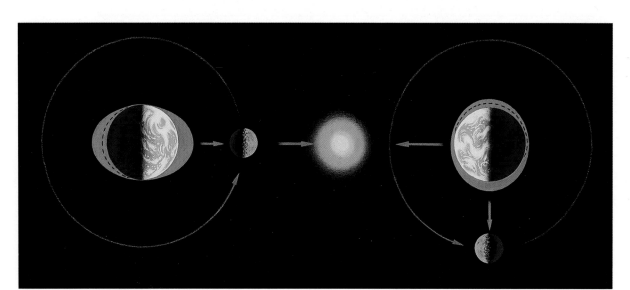

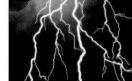

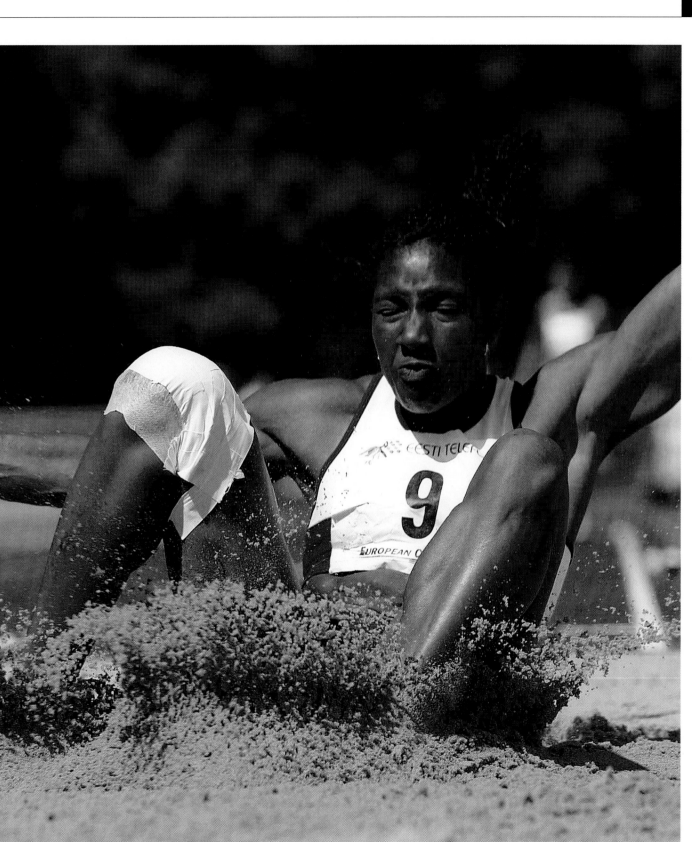

Gravity's many shapes

The force pulls moving objects into graceful curves. That's because vertical and horizontal components of motion act independently.

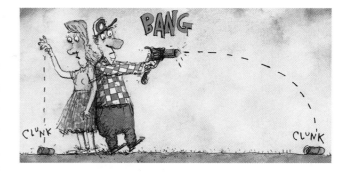

ntil the 17th century most folks figured, as Aristotle had, that heavy things fall faster than light ones. It's common sense: A rock beats a feather every time. We now know that the reason for this difference is air resistance, not gravity. Both the rock and the feather fall at exactly the same rate in a vacuum. But wait! Didn't Newton insist that larger masses generated larger gravitational forces?

That's true. But it's also true that objects with more mass have commensurately greater inertia, and therefore it takes more moxie to get them moving. You can see this the next time you drop the turkey and the gravy boat at once: They hit the dining room rug with a simultaneous splat. Although gravity pulls much harder on the turkey, it also has to overcome the bird's greater resistance to being moved; and while it tugs less strongly on the small gravy boat, it takes far less force to affect its modest inertia. The two effects perfectly offset each other, both objects accelerate equally, and thus they hit the floor at the same time. Galileo Galilei (1564-1642) conducted similar experiments with falling bodies before Newton—though not with turkeys. ■

▲ **Dead heat**
Objects fall at the same rate whether you drop them or fling them out level. A bullet shot from a gun and a bullet dropped from the same height at the moment of firing will hit the ground at the same instant.

▼ **Getting gravity to hold things up**
Linking Manhattan to New Jersey, the George Washington Bridge resists gravity with a typical suspension design, which supports the span's weight on steel cables strung between vertical supports.

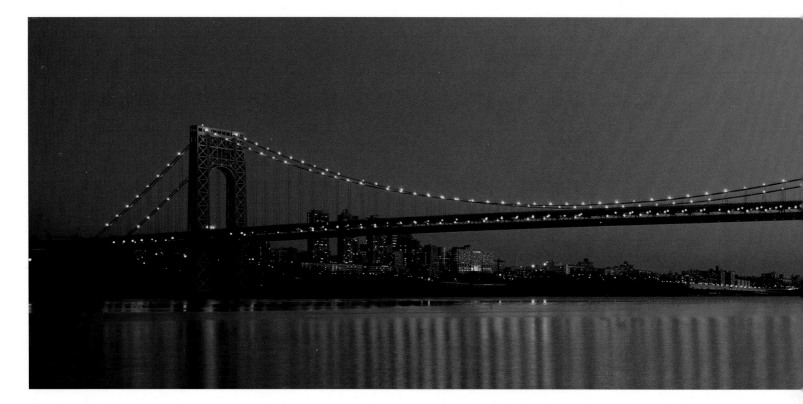

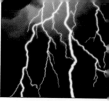

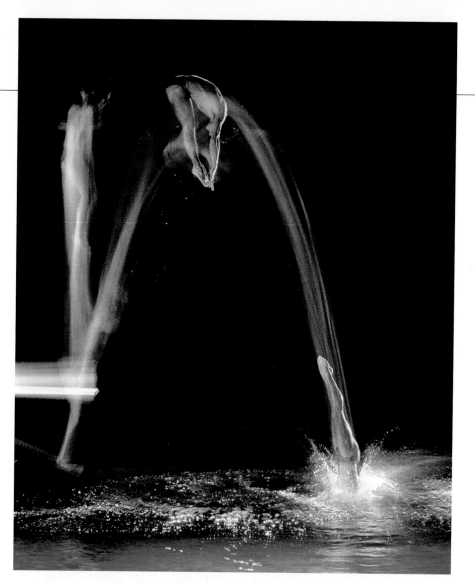

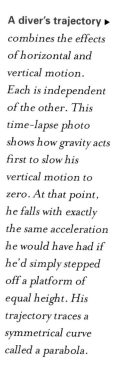

A diver's trajectory ▶ combines the effects of horizontal and vertical motion. Each is independent of the other. This time-lapse photo shows how gravity acts first to slow his vertical motion to zero. At that point, he falls with exactly the same acceleration he would have had if he'd simply stepped off a platform of equal height. His trajectory traces a symmetrical curve called a parabola.

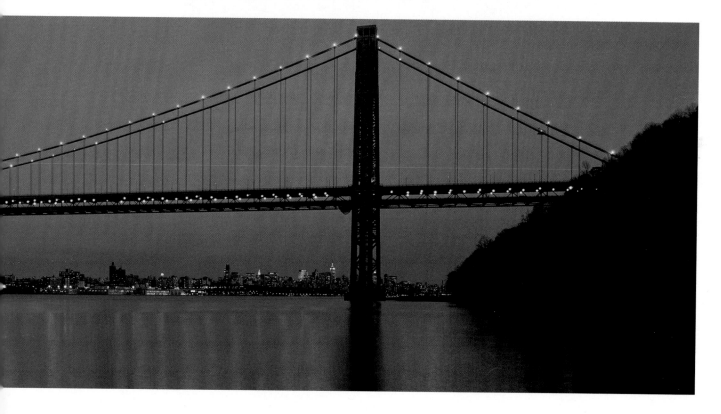

Attraction from afar

Magnetism, the same force that pulls compass needles ever northward, enables credit cards to function—and may even help some animals navigate.

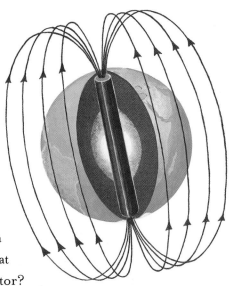

Just as electrical charges can be positive or negative, magnetic fields have two polarities—conventionally called north and south—that attract their opposites and repel their own kind. This similarity with electricity shouldn't surprise, since magnetism is the flip side of electricity. Run an electrical current through a wire and it induces a magnetic field; an even stronger field can be generated by wrapping the wire into a coil and filling the center with metal. That principle is obvious to anyone who's seen a big electromagnet lift a car off the ground. But what about those cute little magnets stuck to your refrigerator? Unless you've got some very odd ones, there are no wires coming out of them. Where's the current?

Actually, it's in the magnet's atoms. Electrons and protons are always in motion. Since any moving charge generates a magnetic field, each electron and proton has one. Usually these fields are oriented haphazardly, canceling each other out. But in materials such as iron, they tend to line up within specific regions called domains. When ferrous rock is molten, all the domains align with Earth's own magnetic field. That's where lodestones and other natural magnets get their overpowering allure. ∎

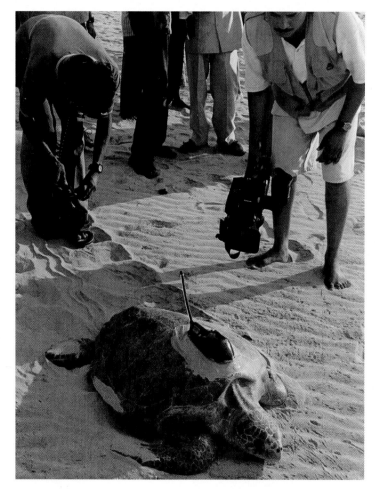

Magnets sealed in rock ▶
Every few hundred thousand years, the Earth's magnetic field reverses. This provides geologists a dandy dating service; since the magnetic orientation of cooling lava always lines up with the prevailing magnetic field. In effect, this stamps them with an eruption date.

◀ Magnet Earth
In this simplified depiction, our planet acts like a huge bar magnet that is about 11 degrees out of line with the axis of spin. Deep in the core, huge blobs of molten iron circulate—driven most likely by convection—forming a titanic dynamo that generates a current, which in turn produces Earth's magnetic field.

◀ Living compasses?
Scientists have found evidence that certain animals such as homing pigeons and sea turtles may be able to use Earth's magnetic field to navigate over substantial distances.

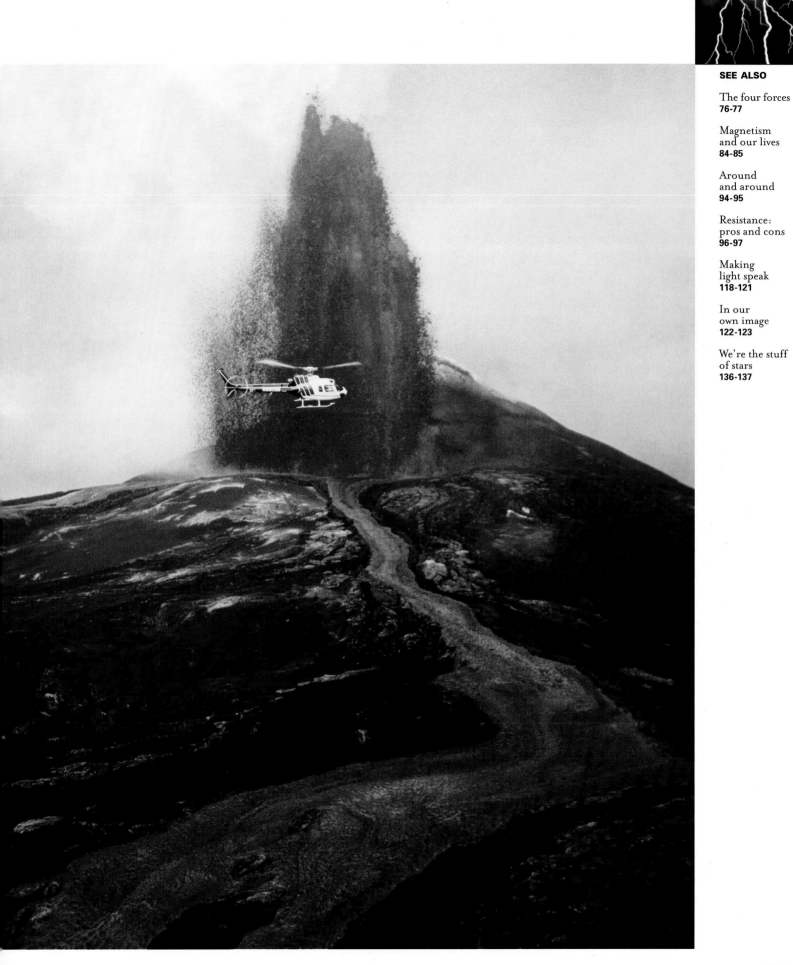

Magnetism and our lives

Far more than a mere oddity of nature, magnetism not only powers the images on your television screen—it makes life itself possible.

I f asked to name all the things responsible for life on Earth, you probably wouldn't think of magnetism. Yet the planet's magnetic field serves a vital function as an umbrella-like shield against the solar wind—a blast of charged particles that streams incessantly off the face of the sun and would wreak molecular havoc on all plants and animals if it reached Earth's surface.

But it doesn't, owing to a very convenient fact: Charged particles from the sun distort Earth's magnetic field, compressing it on the sun's side and dragging it out on the other. Thus most of the supersonic torrent of ions, protons, and electrons billowing toward us is deflected when it hits the zone around Earth called the magnetosphere. Organisms at ground level don't get zapped, and life goes on.

The same principle that deflects the solar wind around Earth puts the picture on tube-type television screens. High-speed electrons fired from electronic "guns" at the back of the set are aimed by magnetic fields so that they slam into little dots of fluorescent phosphors arranged in hundreds of horizontal rows on the inside surface of the picture tube. Magnetic deflector coils—one for each color—continuously vary their field strength, which bends the electron beams slightly so that they sweep from dot to dot across a horizontal row, then drop down to "paint" the next row. When an electron strikes a dot, it so excites the phosphor's atoms that they give off red, green, or blue light. To create the illusion of motion, the dot pattern changes about 30 times per second—fast enough that the human brain perceives it as smooth, seamless action. ∎

EMFs and you

Man-made electromagnetic fields, or EMFs, are very different from natural ones. Earth's magnetic field derives from direct current and doesn't constantly change alignment. But alternating current in your home reverses direction 60 times a second. Every time it does, its magnetic field changes polarity, inducing a current in nearby conductors—including you. Modern life typically requires us to conduct nearly all our daily affairs in a sort of tepid sea of such weak alternating fields.

Magnetic shield ▶
The normal "solar wind" of charged particles is only part of the sun's output. Sometimes enormous eruptions called "coronal mass ejections" hurl a billion tons of hot ions toward the Earth. After a trip of three or four days, they slam into our magnetic shield, or "magnetosphere," with horrendous results. The particles ruin satellites in orbit and induce stupendous spikes of current, blowing out power lines.

▼ Magnets in your pocket
Information is encoded on credit cards, computer disks, and hotel-room key cards by embedding magnetic patterns on very thin layers of material.

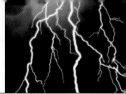

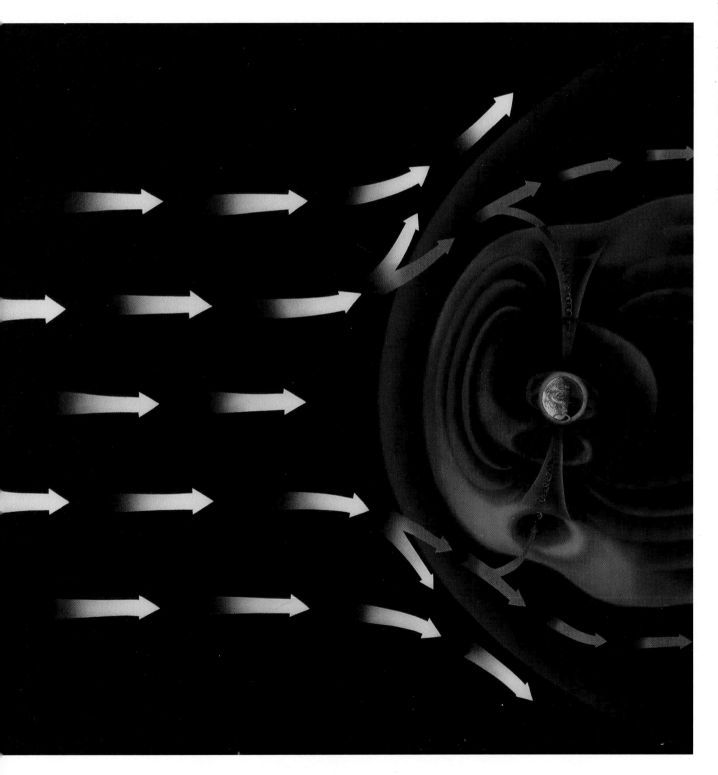

Getting charged

Perhaps the most shocking thing about electricity is that, given the size of electrons, nature gets so much out of so little.

The atomic particles that carry electrical charges are almost unimaginably small: The positively charged proton is about 10^{-15} meter in diameter, or 40 millionths of a billionth of an inch, while the negatively charged electron is at least a thousand times smaller! Tiny vehicles, indeed, for conveying a force. Yet there are so many of them that, in the aggregate, the force between their equal but opposite charges is strong enough to hurl a lightning bolt for miles—or to launch the eerie plumes of an aurora across the Arctic sky, caused when electrically charged particles from the sun hit Earth's upper atmosphere.

Static electricity consists of charges that don't go anywhere. In nature, such charges usually dissipate quickly, often by bleeding off into the air, which absorbs small amounts of electricity if it is suitably moist. In the home, static electricity problems occur most commonly during winter, when air is drier. But when rubbing builds up huge excesses of charge, as happens when your feet scuff across a carpet or atmospheric molecules slide against each other prior to a thunderstorm, the result can be a sudden snap of current—between your hand and a doorknob, or between a storm cloud and your chimney. ■

Light fantastic ▶
One of the upper layers of the atmosphere (ranging from about 50 to 150 miles high) is called the ionosphere because many of its constituent atoms are split into ions by the high-energy particles of the solar wind. Those hyper-excited molecules of the ionosphere shed part of their energy in the form of visible light. Most of the action occurs in bands or ovals within 20 and 30 degrees of Earth's magnetic poles.

Inside the atoms inside matter

Most of what we think of as "solid" matter is nothing but empty space. That's because every atom's nucleus—composed of positively charged protons and electrically neutral neutrons—is relatively distant from the electrons racing around it. If an atom were the size of a football stadium, the electrons would be buzzing around the outermost seats, while the nucleus would be the size of a bee on the 50-yard line. The remaining volume is unoccupied. So when your finger turns the page of this book, why doesn't it just pass right through? The answer is that the electrical forces in the molecules of your hand act on the corresponding forces of the molecules in the page. As a result, things behave as if they were really solid.

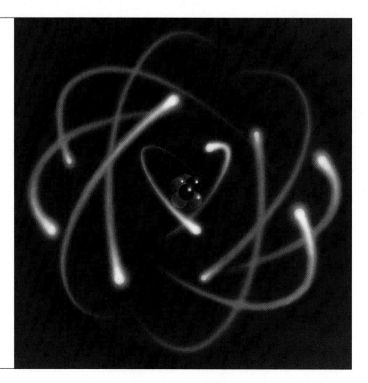

Opposite charges ▶
attract in proportion to the magnitude of each charge and inversely to the square of the distance.

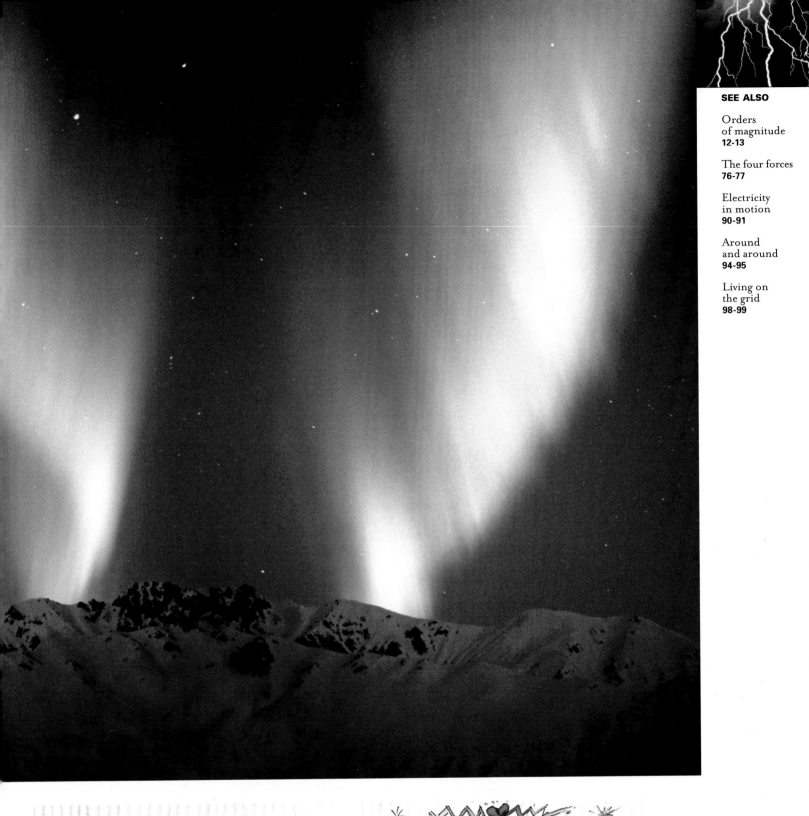

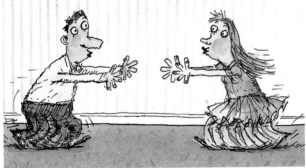

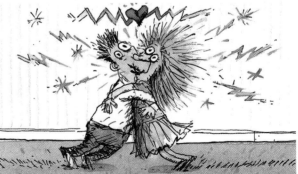

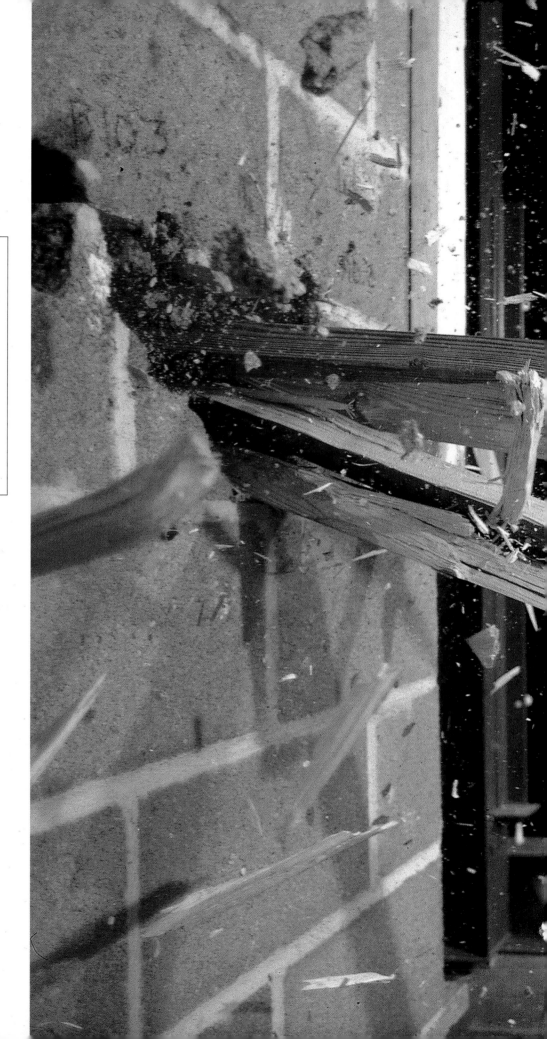

Solidarity

Most of the time, the world seems built of pretty sturdy stuff. To punch an ordinary 2 x 4 through this concrete-block wall, for example, tornado-damage researchers needed a compressed-air cannon that got the wood moving at about a hundred miles an hour. Yet as we've already seen, the atoms that make up wood, walls, and all other matter consist mostly of empty space. The apparent solidity of everything around us—including ourselves—is an enduring testament to the powerful forces that keep atoms and molecules together.

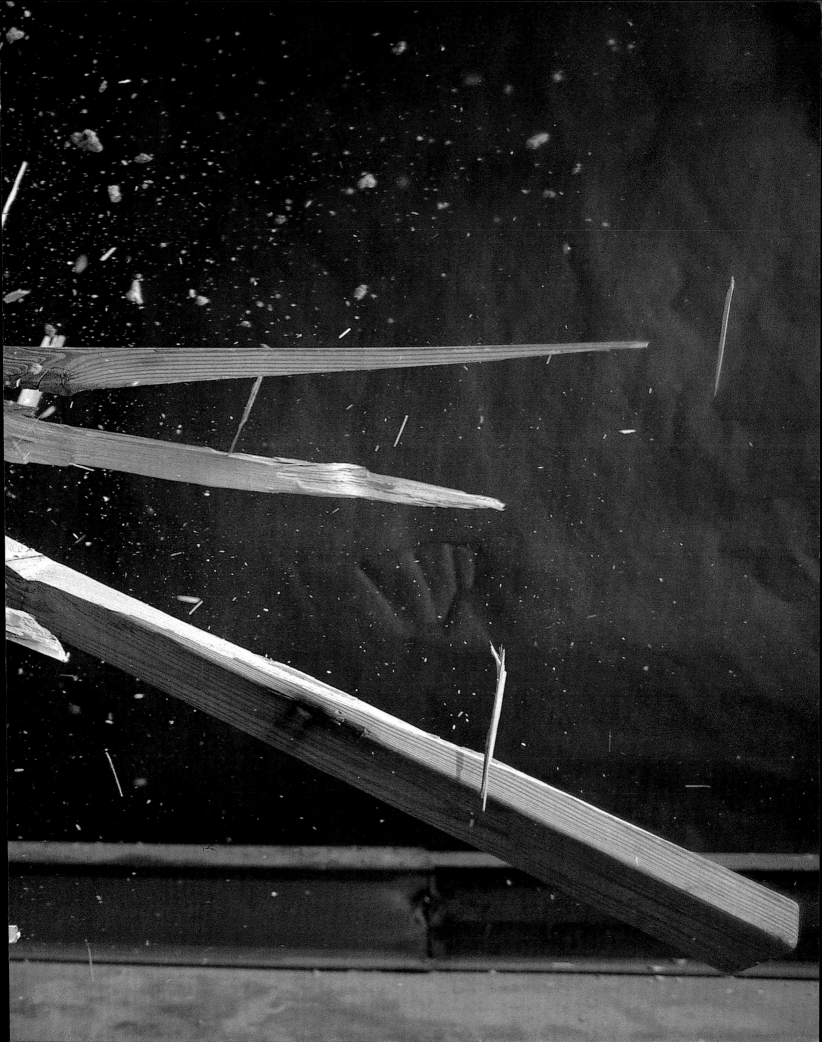

Electricity in motion

When electric charges—positive or negative—travel through a medium, an electric current results. Electrons, being more mobile, usually do the moving.

Current, whether it's direct—as in your car battery—or alternating—as in your house and all its appliances—can be understood most easily by comparing it to water flow. Getting water to your kitchen sink requires three things: a suitable volume of water, a pipe to carry it, and a pressure difference between the source and your faucet. Getting electricity to your toaster works much the same way. First, you need a current of charged particles, that is, a quantity of moving electrons, measured in units called amperes. You also need a material that will properly convey this flow of electrical charges—a wire or some other conducting medium. Finally, you must have a "pressure difference" between the power station and your wall outlet to make the current move in that direction. Such "pressure" is called electrical potential and is expressed in units called volts.

In both plumbing and electrical systems, the quality of the conductor is very important. A water pipe too thin-walled for the volume or pressure forced through it will burst. Similarly, a wire too small for its current will get dangerously hot because even the best conductors, such as copper or gold, offer some resistance to an electrical current. That is, their atoms convert a part of the electrical energy into random motion, or heat. A hair dryer takes advantage of that principle. ■

PUMP (POWER SOURCE)

WATER PRESSURE (VOLTAGE)

WATER FLOW (AMPERAGE)

PIPE (WIRE)

Getting the right flow ▶
Just as the strength of the water squirting out of your garden hose is determined by its quantity and pressure, electrical power (watts) is the product of current (amperes) and potential (volts). Most 120-volt home circuits in the U.S. are designed to handle 15 or 20 amps before they get hot enough to blow a fuse or trip a circuit breaker.

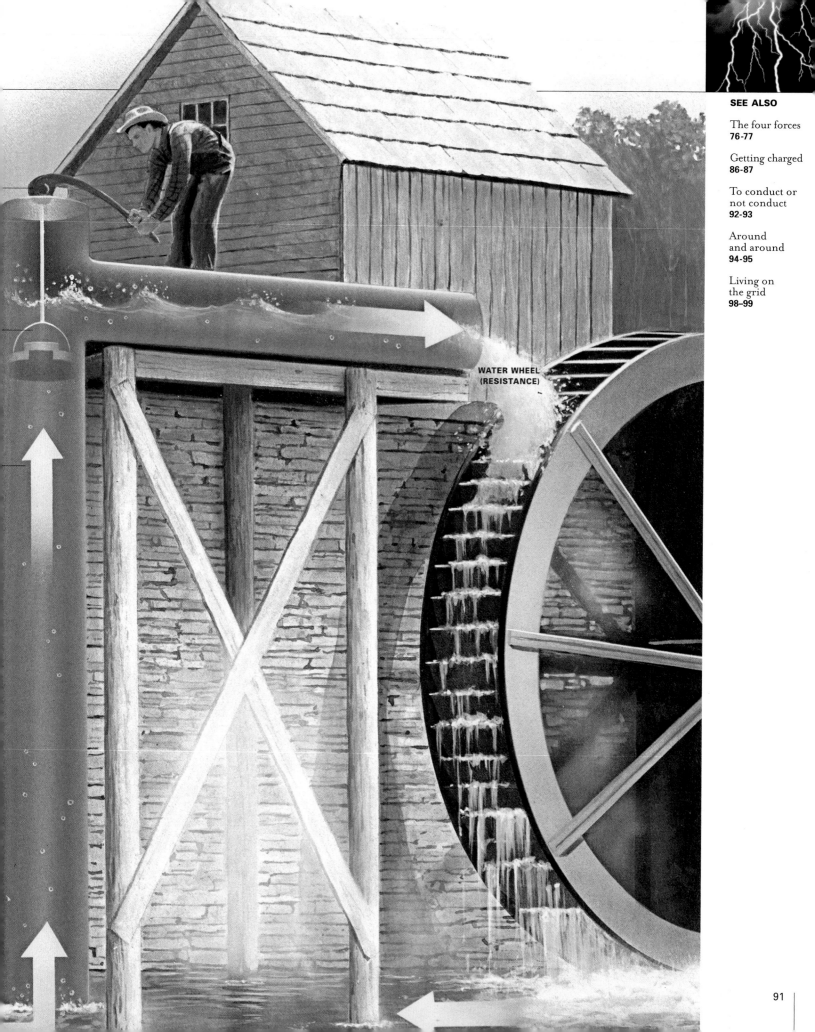

WATER WHEEL
(RESISTANCE)

To conduct or not to conduct

Getting electrons to move is a lot like getting traffic to flow smoothly: It takes the right kind of road system, built of conductors rather than insulators.

The best conductors—copper, silver, and gold—make ideal highways because their atoms form into neatly arranged, tightly packed crystals. Most metals share this orderly structure. But insulators such as rubber or glass are the electron-transport equivalent of wilderness, with disorderly molecular arrangements that tend to deny electrons accessible passage. Then there's a third class of materials known as semiconductors. Like rough dirt roads through the outback, they allow some traffic to get through, but only in certain conditions.

In the atoms of highly conductive solids, electrons in an outer layer called the valence band can scoot around so freely that they are effectively shared by all atoms in common. This helps give metals their distinctive strength. But there is a level just above the valence band where it is even easier for electrons to move. This atomic superhighway is called the conduction band, and an electron needs extra energy to get there; the amount depends on the type of atom. In copper, it's just a quick hop from the valence to the conduction band. So when additional energy arrives, copper's electrons shoot right up into the fast lane. In insulators, however, the distance is so great that there's almost never enough energy to make the jump. ∎

Computer chips

Manufacturers of computer chips engrave silicon wafers with incredibly narrow conductive lines, squeezing millions of circuits onto every tiny chip. These tiny conductors only a few ten-millionths of a meter wide act just like wires running to lights in your home. Just as you can alter the lighting pattern in your rooms by turning lights on or off, the computer chips record and transfer patterns of information by rearranging various currents and charges in microscopic circuits. The currents needed on this scale are so small that mechanical switches would be too big. Instead, tiny electrical switches known as semiconductor junctions are used, enabling designers to pack in more circuits and create highly miniaturized logic systems on little more than the head of a pin.

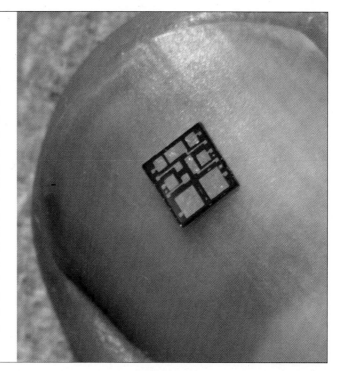

Go with the flow ▶
On a large scale, toll booths act like on-off switches. On the microscopic scale, semiconductors convey current only when you excite them in some way—usually by applying heat, light, or a voltage. Make a sandwich out of two silicon layers— one chemically treated to give it a positive charge, one with a negative charge—and you have a potential current waiting to flow, but prevented from doing so because silicon is such a reluctant conductor. Apply a small voltage between them, using a third piece of silicon or a strip of metal, and you overcome the material's natural resistance and cause a current to flow. That's what transistors do: They're electrical switches as small as one-tenth of a micron.

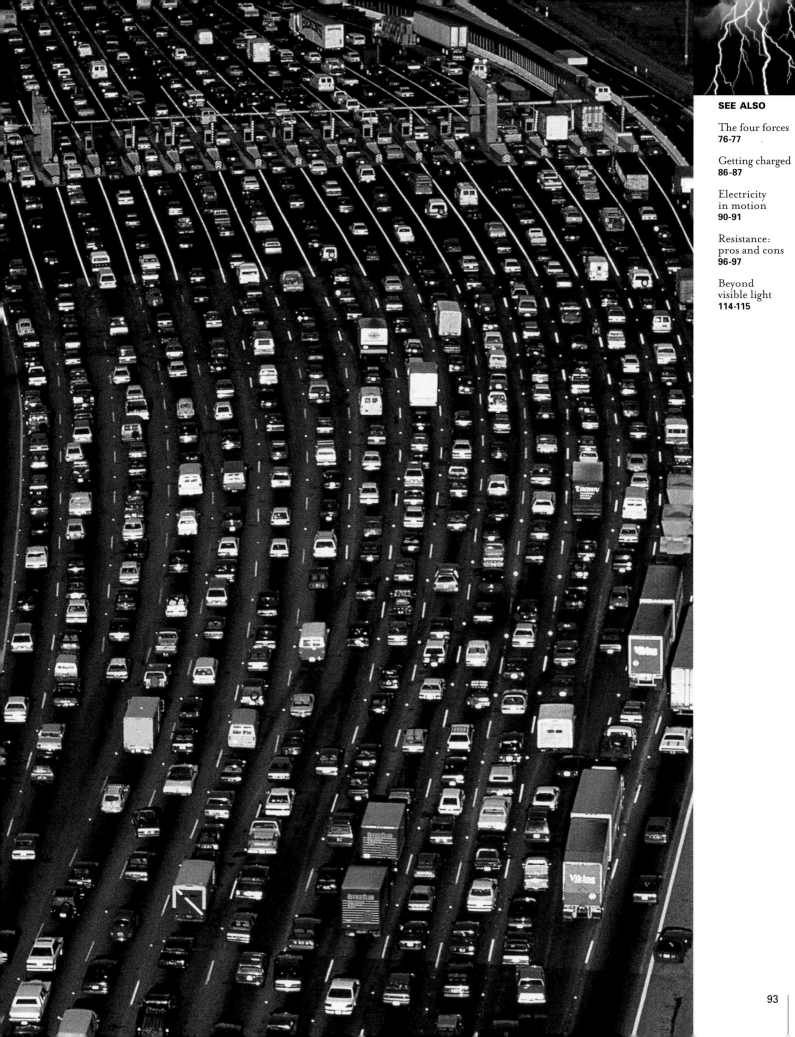

Around and around

Motors and generators are flip sides of the same phenomenon: One turns motion into electricity, while the other transforms electricity into motion.

Many of nature's best tricks are reversible. Heat can make light, and light can make heat. Add potential energy to a pendulum bob by elevating it, and you get kinetic energy back when it swings down. You can raise the temperature of a gas to increase its pressure, or squeeze it to make it hotter. Similarly, electric motors and generators are mirror images of each other.

In generators such as your car's or your power company's, a conductor is rotated through a magnetic field, inducing a current to flow. That generated current then is used to power lights or other appliances. Take similar hardware and simply reverse the order of events, and you have a motor such as the one in your kitchen fan: Here, a current is made to pass through a magnetic field, and the field exerts a sideways force that turns the fan blades. Both mechanisms involve a conductor rotating in a fixed magnetic field. In fact, some electric cars use a combination motor-generator on each wheel. Need to speed up? The batteries discharge, sending current to turn the motors that push the wheels around. But when you brake, the apparatus becomes a generator—scavenging the rotary force of each revolving wheel to generate electricity that recharges the batteries. ■

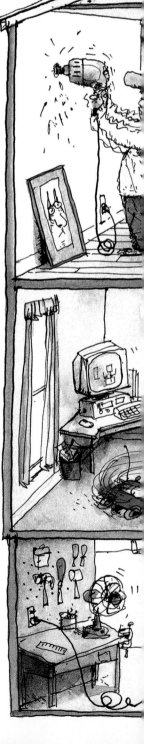

Motors, motors everywhere ▶
What's a modern home without a refrigerator, vacuum cleaner, washer, dryer, or home workshop? America's enduring passion for these and other motorized conveniences has become part of our national identity.

◀ It's all in the turning
In a generator, such as the ones in these windmills, a coil is mechanically rotated within the magnetic field, which induces a current to move through the coil. Because the coil alternately passes both magnetic poles, the induced current flows first in one direction, then the other. The result is alternating current.

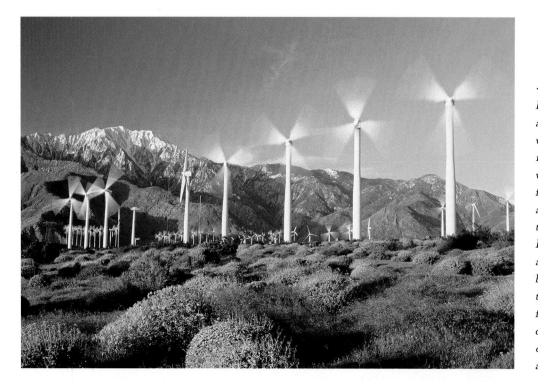

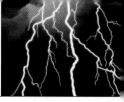

Resistance: pros and cons

In our culture, "resistance" often has negative connotations. But electrical resistance can be a plus: After all, it's why dryers and space heaters work.

ate for work and sprinting to your building? You'll get there a lot quicker if the sidewalk is relatively empty. If it's full of baby carriages, hopscotch games, potholes, and panhandlers, you'll have to swerve or bump into them, which will slow you down. The same thing happens to electrons hustling down a wire. Good conductors allow them to pass virtually unimpeded; high-resistance materials are full of obstacles that electrons bang into, converting some of their energy into random molecular motion, or heat.

In everyday life, resistance eats up as much as a tenth of the electrical

▼ **Busy night**
The locations of cities and transportation corridors are easily seen at night from their stunning illumination in this satellite image. Note that the U.S. interstate highway system appears as a network of glowing threads. That bright line in the northeast corner of Africa is the Nile River shore.

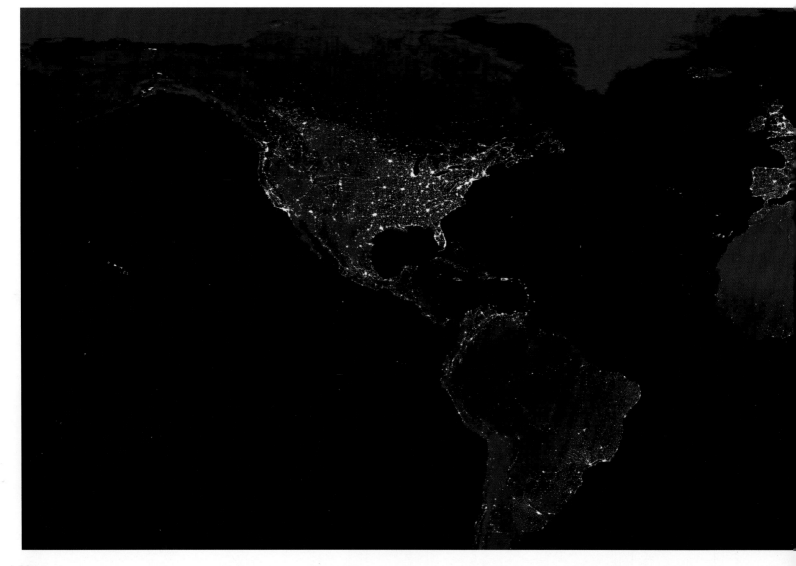

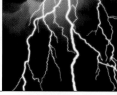

power transmitted in the United States. But it's also an asset. Forcing currents through high-resistance substances makes them furiously hot, giving us electric ovens, clothes dryers, and space heaters. In some resistors, such as the carbon element of an early Edison lamp, the heat is intense enough to create light. The same is true of modern incandescent lightbulbs, which heat tungsten filaments to around 4,000° Fahrenheit. Because so much of their energy is lost as heat, incandescent technology is fast giving way to fluorescent illumination.

In fluorescent lighting, a sealed tube holds an inert gas (such as neon or argon) and a tiny amount of mercury. A device at the tube's terminals stores up electrical charge until the switch is flipped, blasting electrons through the mercury and exciting the mercury vapor to the point at which it gives off ultraviolet radiation. That UV hits a chemical powder coating of phosphors on the inside of the tube, transferring its energy to the phosphors, which in turn sheds that extra energy in the form of visible light. It's a bright idea that's also very cool. ■

▼ **Hot spots**
The pattern of highly illuminated areas reveals the population density on Japan's islands.

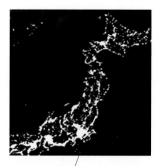

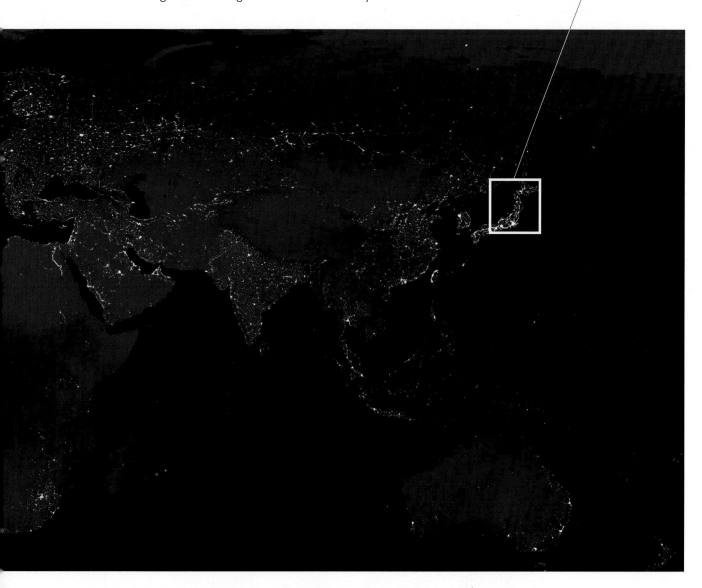

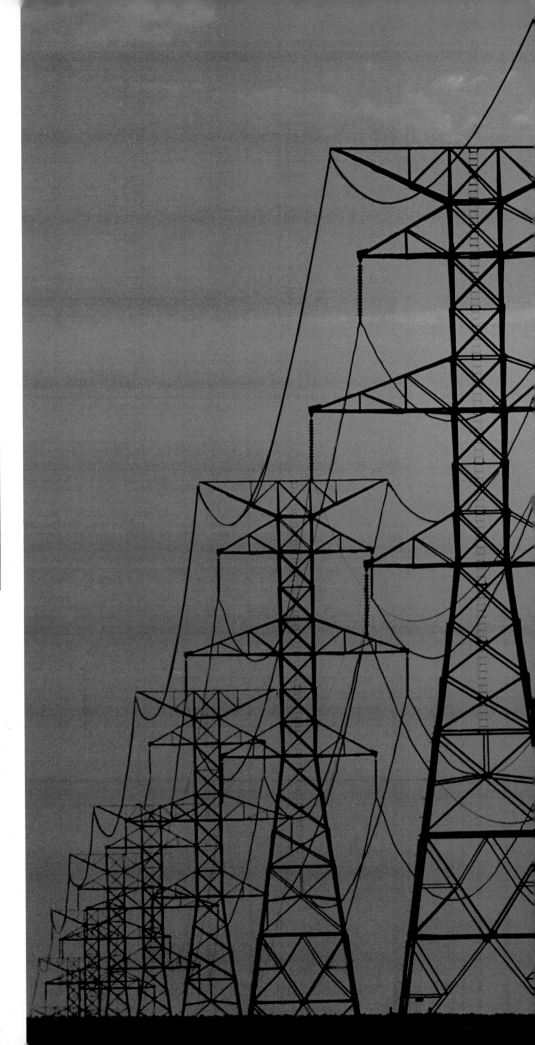

Living on the grid

Ferrying electricity back and forth, transmission lines normally carry from 230,000 to 765,000 volts. They are part of a mighty web that interconnects thousands of electrical utilities with millions of consumers throughout the 48 contiguous states and Canada. Four major regional subdivisions are called Interconnections, or "grids." Each grid is a synchronized, multibillion-watt network. Between power plants and users, countless transformers variously "step up" or "step down" the juice, each time trading voltage for amperage—or vice versa. In North America, electricity is typically generated at 13,000 to 24,000 volts, then is stepped up for efficient, long-distance transmission, then stepped down to under 34,000 volts for local utilities—before being further dropped to the roughly 240-volt service that enters your house.

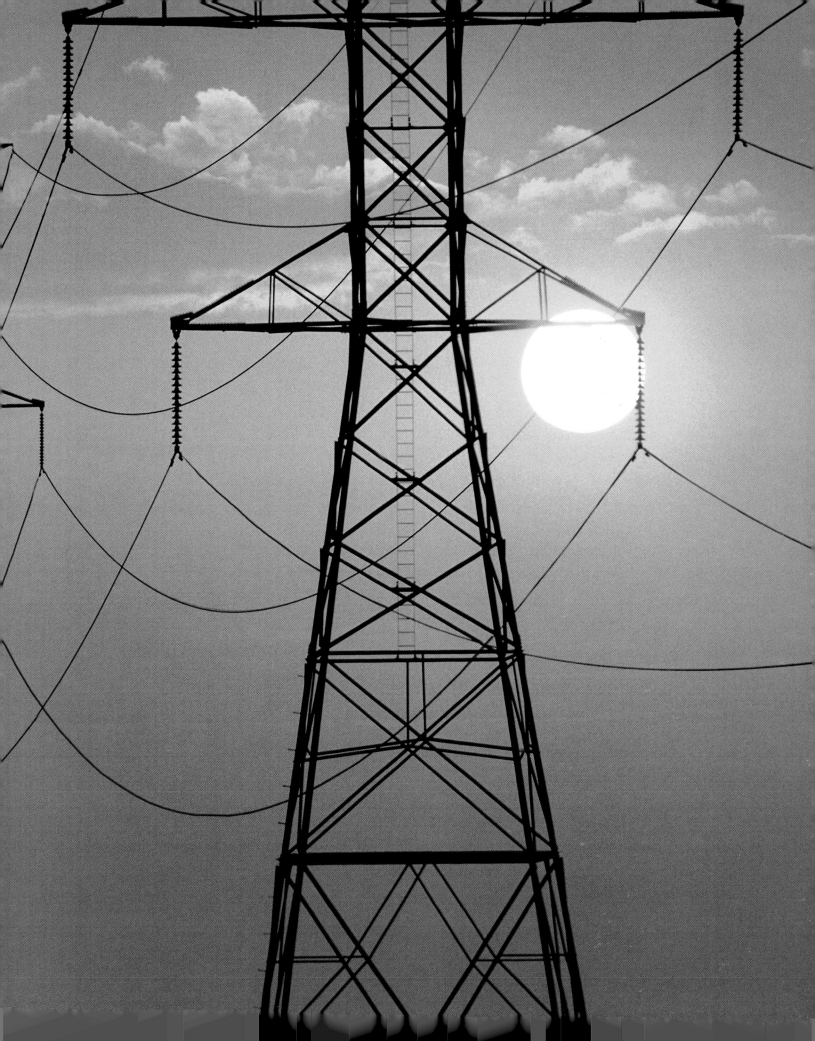

Now you see it...

Light can be visible or invisible, depending on its wavelength. Either way, it's all part of the electromagnetic spectrum.

O f all nature's off-the-cuff miracles, none is so elaborately versatile as the interplay between electric and magnetic fields that we call light—or more properly, electromagnetic waves. They are created whenever an electrical charge is accelerated—that is, changes its speed or its direction—producing a varying electric field that sets up a varying magnetic field that in turn prompts another electric field, and so on. The result is a pair of symmetrically entwined electrical and magnetic fields that oscillate in planes at right angles to one another, engendering and sustaining each other as they travel through space at 186,000 miles per second, the universal speed limit we call the speed of light.

Visible light, however, is only a tiny slice of the electromagnetic spectrum: What you see is far less than what you get. The spectrum consists of a continuum of infinite variety, from low-frequency radio waves hundreds of thousands of feet long to high-frequency x-rays and gamma rays with wavelengths around a trillionth of an inch! The higher the frequency (and the shorter the wavelength), the more energy each wave carries. Thus you can't even feel a TV signal, but infrared waves can warm your skin and enough x-rays can actually damage the DNA within cells.

One of the paramount triumphs of human progress is our ability to "see" in other wavelengths than visible light, which ranges from 400 (very blue) to 750 (deep red) nanometers, or billionths of meters. Astronomy, medical research, military technology, and sciences of nearly every sort now routinely use numerous parts of the electromagnetic spectrum to gain knowledge. ∎

The ouch factor ▶
Looking for evidence that shorter wavelengths carry more energy? Check your sunburn, caused by ultraviolet radiation, which packs enough energy to damage, or even kill, sensitive skin cells.

▼ Broad spectrum of uses
Like visible light, every segment of the electromagnetic spectrum can be reflected, bent, or spread out to separate constituent wavelengths. This malleability begets versatility. Microwaves, for example, can heat food, carry surface communications, and signal satellites.

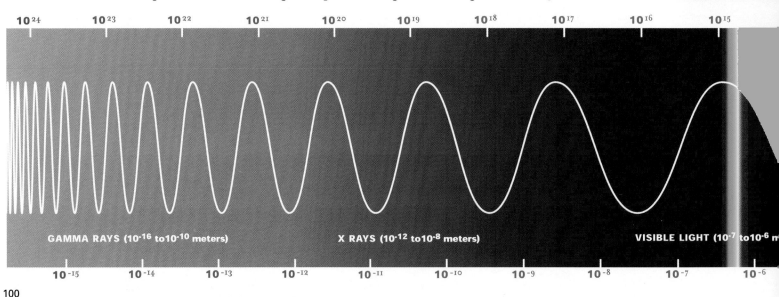

10^{24} 10^{23} 10^{22} 10^{21} 10^{20} 10^{19} 10^{18} 10^{17} 10^{16} 10^{15}

GAMMA RAYS (10^{-16} to 10^{-10} meters)   **X RAYS (10^{-12} to 10^{-8} meters)**   **VISIBLE LIGHT (10^{-7} to 10^{-6} m**

10^{-15} 10^{-14} 10^{-13} 10^{-12} 10^{-11} 10^{-10} 10^{-9} 10^{-8} 10^{-7} 10^{-6}

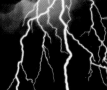

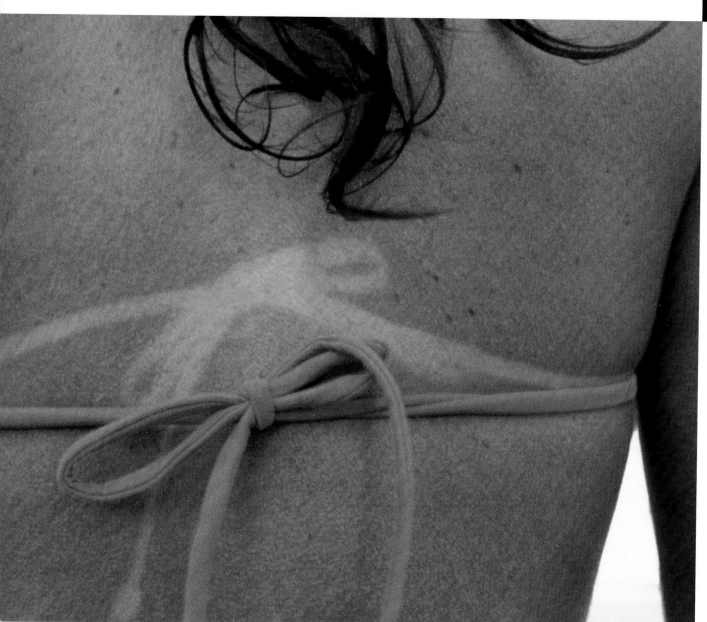

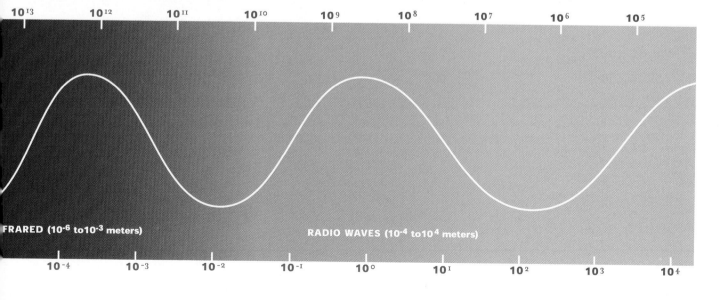

10¹³ 10¹² 10¹¹ 10¹⁰ 10⁹ 10⁸ 10⁷ 10⁶ 10⁵

INFRARED (10⁻⁶ to10⁻³ meters)   RADIO WAVES (10⁻⁴ to10⁴ meters)

10⁻⁴ 10⁻³ 10⁻² 10⁻¹ 10⁰ 10¹ 10² 10³ 10⁴

Light of many colors

Seeing is believing? Not with color, which depends on what's absorbed and what's reflected. That sports car you want is really every color <u>except</u> red.

The wavelengths visible to our eyes constitute only a thin sliver of the entire electromagnetic spectrum, but that happens to be the range our local star specializes in. The sun puts out some 49 percent of its radiation in the visible range, about 46 percent in infrared, and the balance in the ultraviolet. But, come to think of it, if all colors are just part of the same spectrum—and the same sunlight shines on everything—why do we see different colors, anyway? Why isn't everything one big, undifferentiated glow?

As usual, it comes down to atoms. When light hits an object, some of this electromagnetic radiation is absorbed and some is reflected. Atoms can absorb only those waves whose energy content corresponds with possible energy states those atoms can have. Suppose you run a hardware store that has bins for 2-inch, 4-inch, and 6-inch nails. One day somebody delivers a bunch of 3-inch nails. You have nowhere to put them, so you send them back. Similarly, an atom absorbs only those photons that fit, reflecting the rest. Each element has a slightly different set of wavelengths it can absorb. Things that appear largely transparent to our eyes, such as air and water, don't have the right energy levels to absorb visible light of any color.

Unless an object emits its own light, its color is a function of those rays it discards: The blue field in the American flag is that shade because the atoms of chemical dye in the cloth absorb almost every wavelength except blue, particularly a lot of orange. Apples appear red—most of them, anyway—because they absorb blue and green wavelengths. Leaves are green because the pigments involved in photosynthesis happen to be most receptive to red and blue light. ■

Blue skies and pink sunsets ▶

Light waves scatter off air molecules and dust in the atmosphere. But the very short blue waves scatter the most. So when you look up at the sky, you're seeing those waves that have ricocheted toward your eye rather than passing straight through the air. At sunset, when light has to pass through the maximum amount of air to reach your eyes, only the longer-wavelength red and orange rays make it through unscattered.

Bushmasters and bees

Certain kinds of snakes—especially pit vipers, which include rattlesnakes and bushmasters—evolved in caves and other low-light venues where they developed the physiological equipment to sense infrared (that is, heat) radiation. Tiny pits, located either between nostril and eye or within the scales of the lip line, are sensitive to very slight changes in temperature, allowing the snakes to prey upon small, gray, warm mammals amid small, gray, cooler rocks. Military "heat-seeking" missiles use an electronic version of the same technique. Meanwhile, at the opposite end of the visible spectrum, some bees can detect ultraviolet light, a talent that helps them discriminate among different flowers.

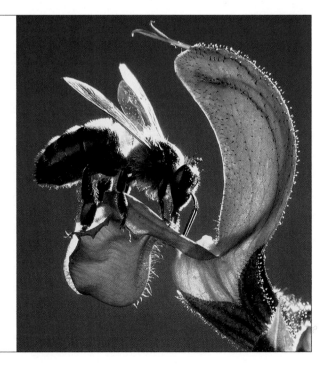

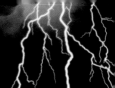

SEE ALSO

Making waves
56-59

Reflection
and refraction
60-61

Now you
see it…
100-101

Managing light
for fun
and profit
104-105

Making
light bend
108-109

Seeing and
believing
246-247

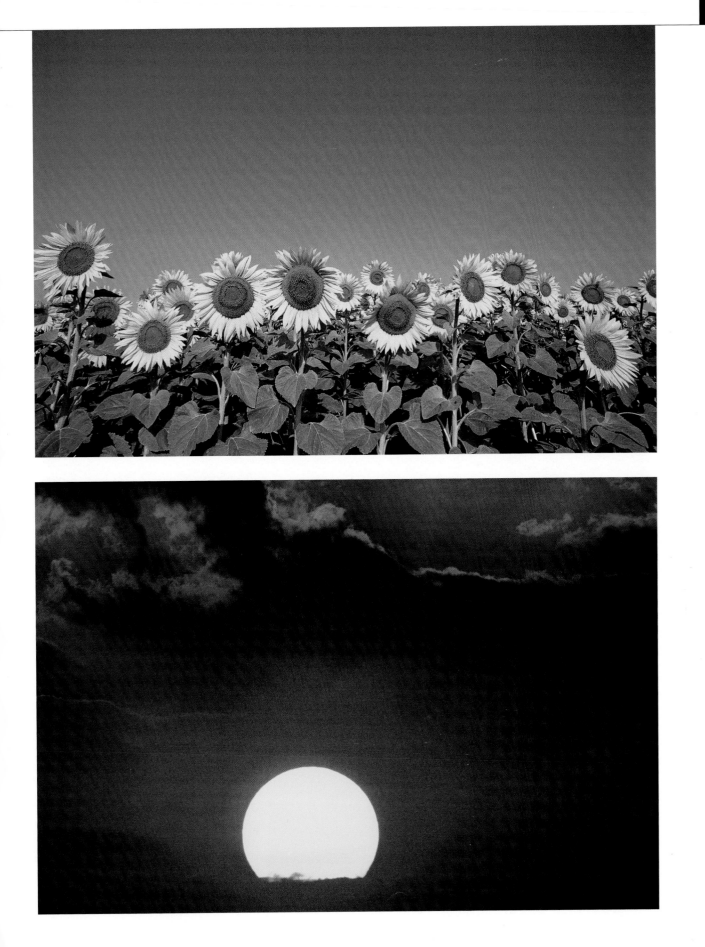

Managing light for fun and profit

Whether you're picking out house paint or a necktie, using binoculars or slapping on sunblock, much of your day is spent trying to control light.

Obviously, the mirrors in a solar collector need to be properly aimed to concentrate sunlight in the right places and thus generate maximum electricity. In this and in many other ways we manipulate electromagnetic waves every day. Whenever we select a tie, we're actually basing our decision on the abilities of its molecules to absorb and reflect various wavelengths of light.

We use a sunscreen because it gives us a thin layer of something nature didn't supply—protective chemicals that soak up ultraviolet radiation before it hits our skin. One highly popular compound is titanium dioxide, a major gleam-enhancing ingredient in a host of products from house paint to cosmetics and toothpaste. It is famous for its ability to reflect and absorb ultraviolet radiation.

Earth, of course, has its own sunscreen in the form of a layer of ozone—an unusual, three-atom compound of oxygen. Created in the stratosphere about 10 to 20 miles up, ozone absorbs much of the ultraviolet radiation that would otherwise reach the planet's surface. Ultraviolet is bad news for living things, since its high-energy waves can destroy or severely mutate DNA, the genetic material essential to all cellular reproduction. A recent, disturbing decline in ozone abundance has been traced to the use of popular refrigerants and solvents collectively called chlorofluorocarbons, or CFCs, which rise into the stratosphere and there break up ozone molecules. International agreements now mandate replacing such chemicals with compounds that are friendlier to the ozone layer. That's why, if you have a new car or home air conditioner, it probably uses a non-CFC gas. ∎

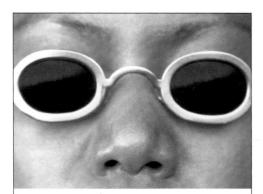

Ultraviolet and the eyes

One danger of ultraviolet is skin cancer; another is cataracts, a whitening of lens proteins similar to what happens when you fry egg white. Sunglasses reduce the number of photons striking your eyes, either by containing pigments that absorb some wavelengths or because they're polarized to admit only a fraction of the incoming waves. Electromagnetic waves oscillate in planes. The waves of ordinary, unpolarized sunlight have planes oriented at random. One way to polarize light is to force it through a barrier with a very narrow slit, so that only waves that are aligned with the slit will get through. Polarizing lenses contain long, thin, parallel molecules that create, in effect, a series of slits—which cut glare because they reduce the reflection caused by randomly oscillating light waves. Certain camera filters do the same kind of thing .

Power of light ▶
Ordinary sunlight contains an enormous amount of energy. Even after it makes its way through the atmosphere, solar radiation is carrying hundreds of watts per square yard of illumination. So engineers have devised ways to focus that energy and convert it to electrical power—either by harnessing its heat or by photoelectric conversion.

The eyes have it ▼
Some animals have obvious adaptations to better manage light. Owls, for example, have evolved extra-large pupils and specially receptive retinas that help them see well even in low-intensity situations.

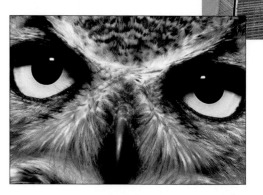

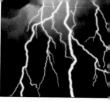

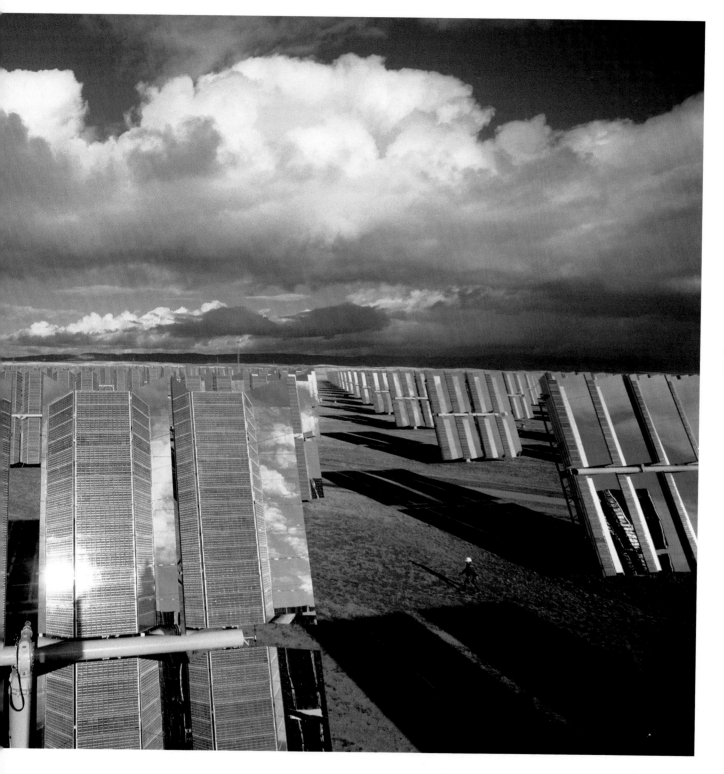

Of sunlight and Earth

Human alteration of the environment has led to a lot of things we never used to worry about, among them the greenhouse effect and the heat-island effect.

s long as Earth has had an atmosphere, it's had some "greenhouse effect"—a phenomenon whereby various gases in the air behave like the glass panes of a greenhouse, letting in light but trapping some of the heat before it can escape back into space. In general, that's good. The planet would be an ice ball without it. But industry and land use are changing the ratios of gases in our atmosphere, prompting worry about increased global warming.

Nitrogen and oxygen—the major constituents of air—are transparent to infrared. But certain gases—particularly water vapor and carbon dioxide—absorb infrared radiation and warm the atmosphere. At present, CO_2 makes up only .03 percent of our air. But it's rising about one part per million per year, an amount that experts believe is sufficient to raise average global temperatures by one to three degrees Fahrenheit early in this century. ∎

Islands of heat ▶
Cities and densely packed suburbs are short on greenery and long on concrete and asphalt. Those artificial substances are far more prone to trap and re-radiate heat from sunlight than are forests and fields. As a result, urban areas often become "heat islands" in which the average high temperature is many degrees hotter than in the surrounding countryside.

◀ City vs. country
This 1988 LANDSAT image of the Miami area shows how the amount of solar radiation reflected from Earth's surface can differ dramatically, depending on whether urban or vegetated areas are involved.

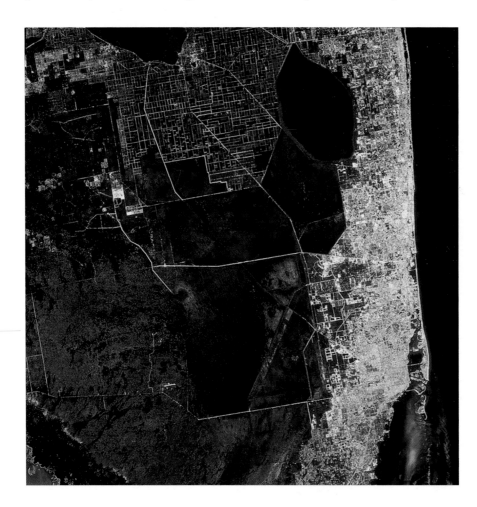

INFRARED EMITTED BY EARTH

INFRARED RE-RADIATED BY ATMOSPHERE

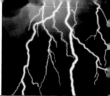

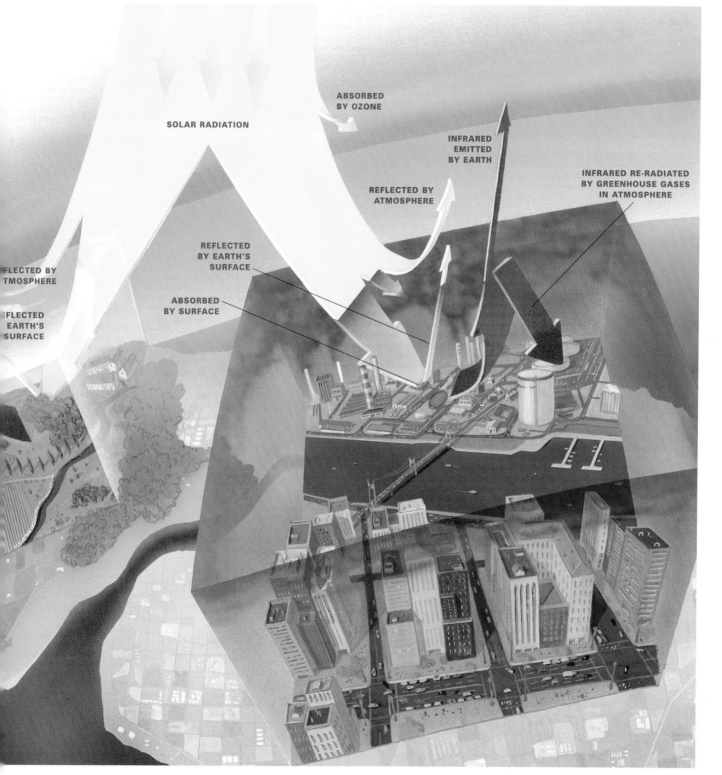

SOLAR RADIATION

ABSORBED
BY OZONE

INFRARED
EMITTED
BY EARTH

REFLECTED BY
ATMOSPHERE

INFRARED RE-RADIATED
BY GREENHOUSE GASES
IN ATMOSPHERE

REFLECTED
BY EARTH'S
SURFACE

FLECTED BY
TMOSPHERE

ABSORBED
BY SURFACE

FLECTED
EARTH'S
SURFACE

Making light bend

Although we often treat the speed of light as a constant, it isn't. The textbook 186,000 miles-per-second figure applies only to light traveling in a vacuum.

n the real world, light's speed varies somewhat according to the material it is moving through, whether that's air, water, glass, or Jell-O. Each time a light wave angles into a different medium, its velocity changes—and the light itself actually bends. That's called refraction, and it's occurring this very moment, as light rays stream off this page and hit your glasses or the lens in your eye.

To picture it, imagine an approaching light wave as a 120-piece marching band with 12 musicians in each row, tramping in time across a parking lot. Suppose that a circle of fresh tar, 20 yards across, just happens to be in their way. Because the tar's border is curved, some front-row musicians will hit it before others. They slow down immediately, while the rest of the row keeps moving at normal speed. The whole line will bend slightly toward the side where the first marcher hit the goo. That's exactly why light waves refract—bend—when they pass through a curved lens or a triangular prism. They always curve toward the thickest part of the glass, because light travels more slowly through glass than through air. ■

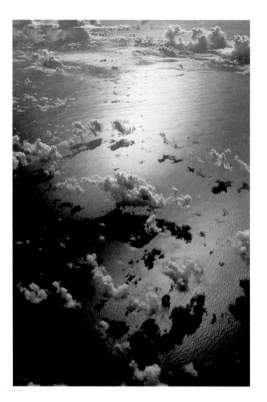

Natural lenses ▶
Although we're usually most familiar with synthetic optics in eyeglasses and contact lenses, any transparent substance curved in the proper proportions can serve to focus light into an image. In this case, an ordinary water drop serves as a lens.

◀ Watery prisms
When sunlight is caught in spherical water droplets and reflected outward again, each wavelength is bent, or refracted, to a slightly different degree. As a result, we see each wavelength separately. That's what makes the rainbow colorful.

Making optical corrections

Lenses of various curvatures compensate for irregularities in our eyes. Sharp vision occurs only when all incoming rays focus, converging precisely on the surface of the retina, the photosensitive lining at the back of the eyeball. If you're nearsighted, your lens actually focuses slightly in front of the retina; you need glasses that spread the rays, moving the focus back. If you're farsighted, the reverse is true, and your glasses should make the rays converge sooner, moving the focus forward.

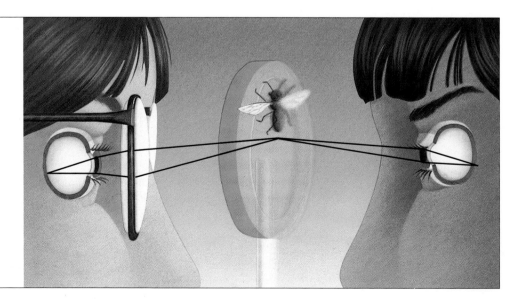

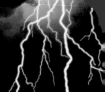

SEE ALSO

Making waves
56-59

Reflection
and refraction
60-61

Now you
see it…
100-101

Light of
many colors
102-103

Focusing on
worlds near
and far
110-113

Seeing and
believing
246-247

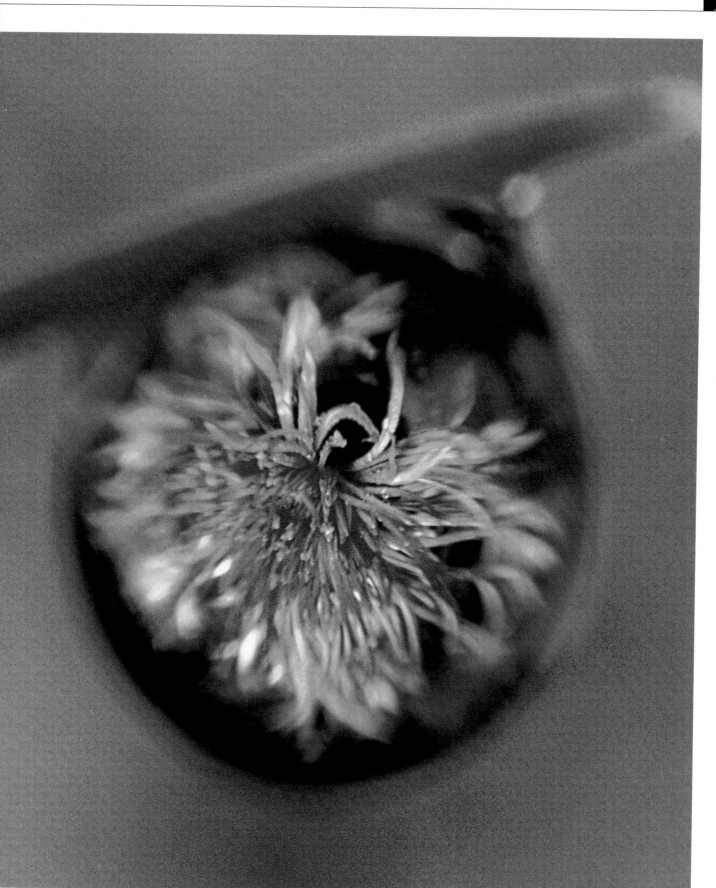

Focusing on worlds near and far

The basic principles of optics empower us to see realms far larger—and smaller—than earlier generations ever dreamed existed.

Sometimes just seeing straight isn't enough, and what really matters is fine detail. That calls for optical magnification—that is, getting a small or distant object to form a larger picture in the eye than it ordinarily would. Once again, the light-bending powers of refraction come to our aid.

A simple magnifying glass, a telephoto camera, a laboratory microscope, and a ten-ton telescope all use different combinations of lenses to achieve the same basic effect. They select only a portion of the light rays from an object and then refract them so that they make a disproportionately large image on the retina or a camera's image plane. The more the rays are bent, the larger the area the magnified image will occupy. For example, when your unaided eye looks at a small pine tree a hundred yards away, it might make an image only a couple of degrees wide—because the rays of light coming from opposite ends of the tree are so close together. But hold up a 40-power telescope, and suddenly the tree fills your field of vision, projected as an arc 70 or 80 degrees wide on the back of your eye. Magnifying power is determined by the ratio of those angles of vision, aided to unaided.

In astronomy, size matters. Ever since Galileo started playing with lenses and rearranging them in different sizes of tubes, astronomers have been hungering for ever bigger lenses and ever larger telescopes. That's because light diminishes according to the square of the distance from its source, and there just isn't much starlight left by the time it gets here from dozens of light-years away. (A light-year is the distance light travels through a perfect vacuum in a year. At 186,000 miles per second that works out to roughly six trillion miles.)

In theory, the larger the lens, the larger the image. But there are limits: A glass lens big enough to capture the sparse rays from some distant stars would sag under its own weight. One solution is to switch to mirrors. Like lenses, mirrors shaped to specific curvatures can dramatically magnify images. They also possess a structural advantage over lenses: Because they only need to reflect light rays, not transmit them, they can make use of steel and other supports. So the world's largest optical telescopes are all the reflecting type. Hawaii's W. M. Keck Observatory boasts a composite mirror 33 feet in diameter. ■

▼ Heads up
Intensely energetic objects such as stars or galaxies emit radiation in many wavelengths besides the visible. So astronomers try to examine them in radio, ultraviolet, x-ray, and even gamma radiation. Some radio telescopes look a lot like this satellite dish. Others are spread across dozens of acres in giant collective arrays.

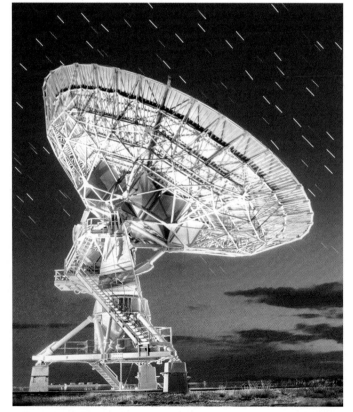

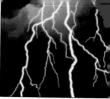

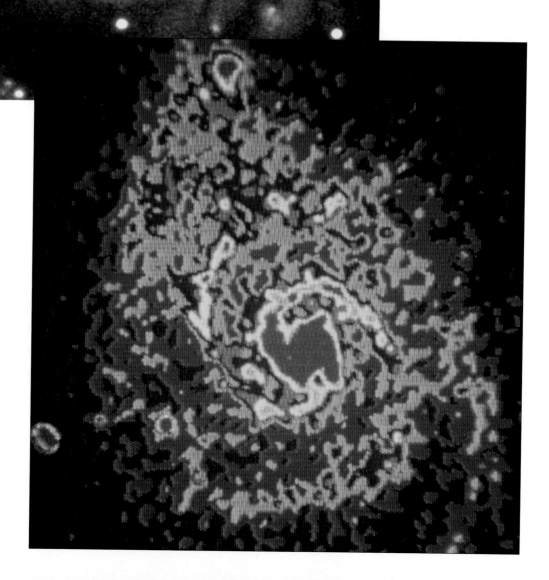

◆ **Viewing past
and future** ▶

*Optical telescopes
enable our eyes to see
distant scenes of
stellar genesis such as
the image above of
spiral galaxy NGC
5194 some 13 million
light-years from
Earth. But viewing
the same object in
radio wavelengths
(right) reveals
additional structure
and enables
astronomers to detect
concentrations of
matter that do not
radiate much light in
the narrow visible-
wavelength range.*

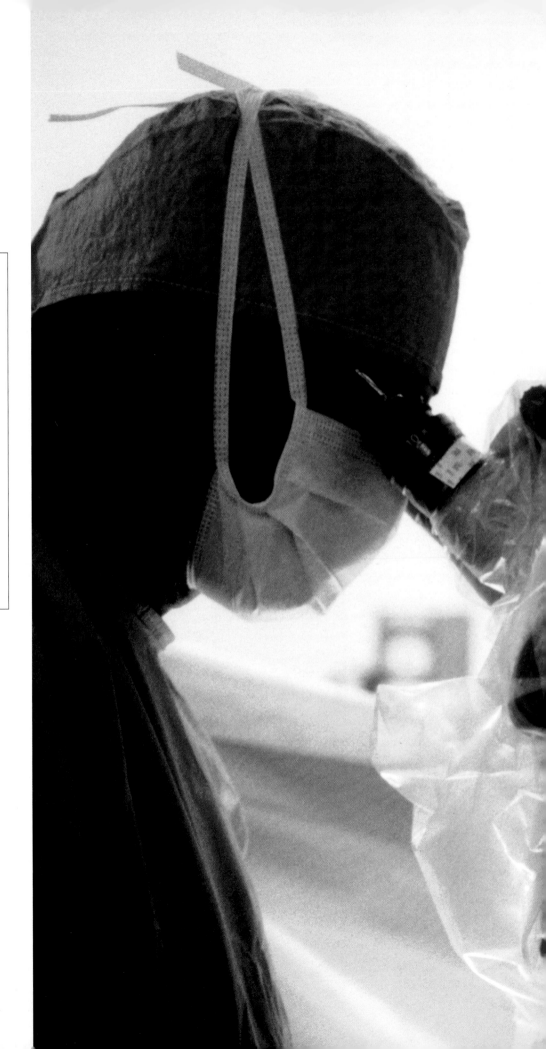

Inner vision

A surgeon performing very delicate forms of surgery can use high-powered microscopes. Sometimes a video camera picks up the image and routes it to a closed-circuit television. The best optical microscopes have magnifications between 500X and 2,000X. Beyond that point, it's hard to get a decent image with visible light. One reason is diffraction: Waves tend to bend around obstacles in their way and scatter somewhat, becoming less focused. Just as light waves diffract, so do sound waves—which is why you can hear a noise from around a corner. Another problem is that we cannot see any detail smaller than the individual rods and cones in our retinas. The spacing between those cells—rarely less than three-millionths of a meter—sets a firm limit on the eye's resolution.

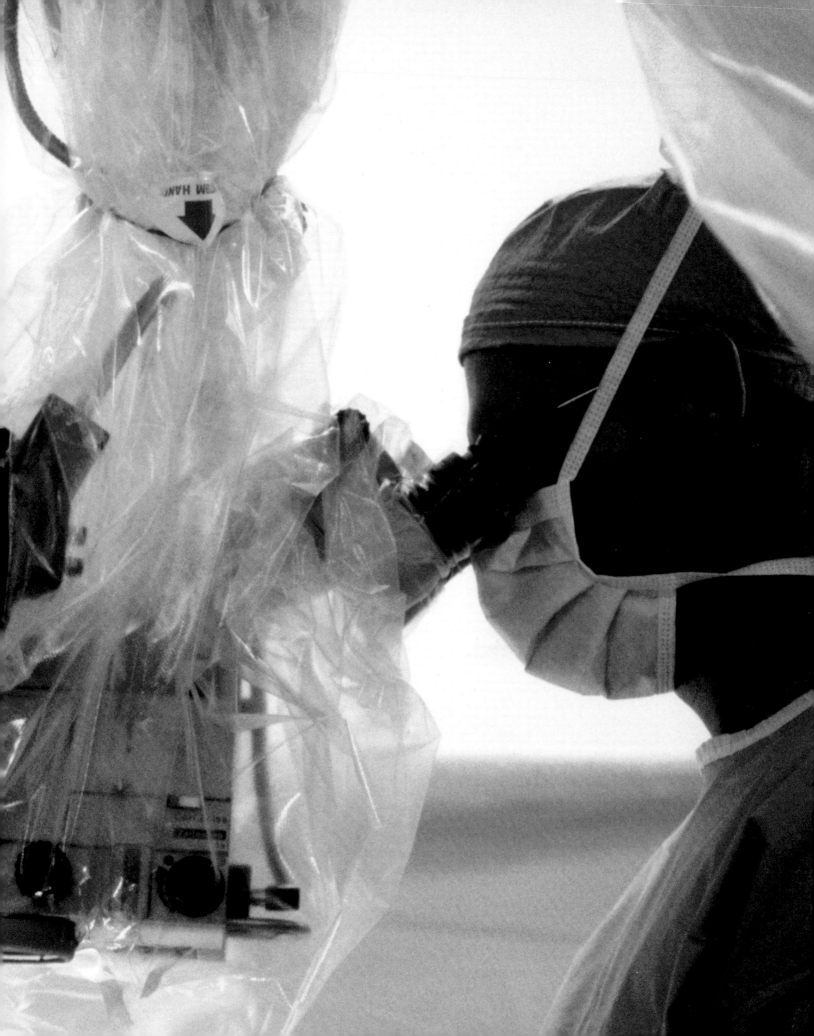

Beyond visible light

There's a whole spectrum of electromagnetic waves out there—and quite a few segments are already being exploited for human use.

Want to break into broadcasting the easy way? Just scuff your feet hard against a nylon rug until your body has a whopping electrical charge. Then hop up and down. Congratulations! That charge you've just accelerated is radiating electromagnetic energy like ripples in a pond. If your hops take a second each, you're transmitting electromagnetic waves at a frequency of one cycle per second, or one hertz. Since the speed of light is 186,000 miles—or about 300 million meters—per second in a vacuum, you're broadcasting at a wavelength of 300 million meters, toward the bottom of the radio band of the electromagnetic spectrum.

Every day, radio and TV stations make electric charges move back and forth in transmission antennas, generating electromagnetic waves from a few inches to several miles long. The signals they send, whether words or pictures or music, are embedded in a unique "carrier" frequency. Radios and television sets detect those waves and then electronically strip away the carrier frequency, leaving only the information content. In general, electromagnetic waves travel in straight lines and so reception is limited to areas within the line of sight of the transmitter. But some shortwave radio signals are the right frequency to reflect off the electrically charged ionosphere layer. As a result, shortwave broadcasts often can be heard thousands of miles away. ■

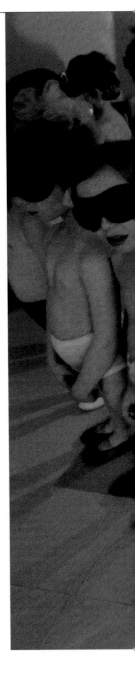

Using ultraviolet ▶
Kids in Murmansk, above the Arctic Circle, get a healthy dose of UV light. Elsewhere, microchip makers use UV to engrave tiny circuits on silicon wafers. They coat the wafers with a UV-sensitive material and shine UV light through stencil-like "masks" cut to the intended patterns. Then a solvent washes away exposed areas, leaving threadlike indentations for electrical contacts.

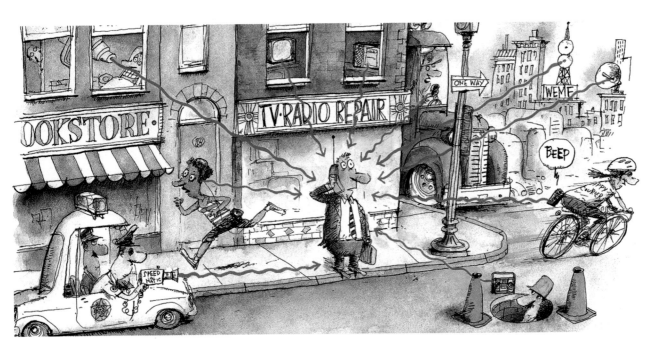

◀ Sharing spectrum
AM radio operates from 530 KHz— thousands of cycles per second—to 1.6 MHz—millions of cycles per second. FM goes from 88 to 108 MHz. Likewise, TV, cell phones, and commercial and military radar all have their special bands.

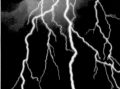

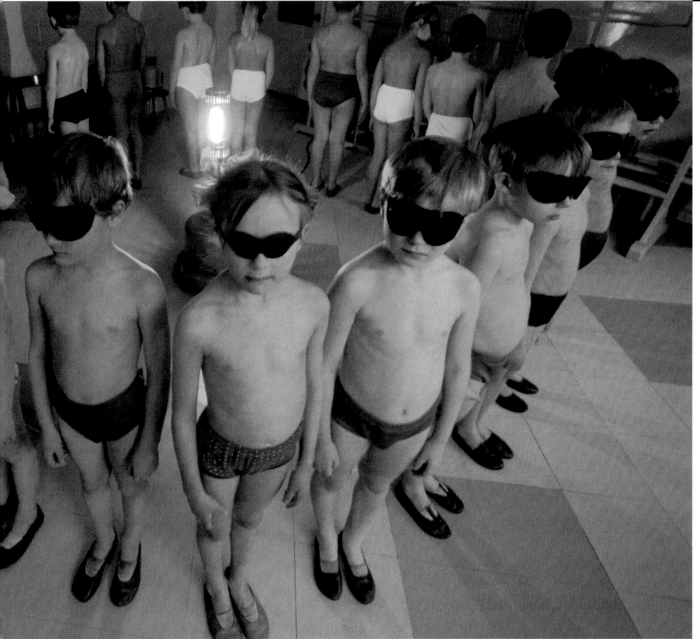

The power of x-rays

Possessing even shorter wavelengths and higher frequencies than ultraviolet light, x-rays are created by bombarding a metal target with a high-speed electron stream. They have so much energy that they easily penetrate soft tissues and are only slightly slowed or deflected by harder ones, such as bones. This difference in penetration shows up as light and dark areas on a photographic plate or film.

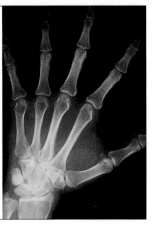

At the cutting edge of light

Once dismissed as science fiction, laser technology has become a growth industry with a host of applications in everything from health care to media.

Getting light waves to cooperate with each other can be harder than herding cats, because most light sources—from a burning match to a fluorescent bulb to the sun—emit a random variety of rays. In wavelength, amplitude, phase, and even in direction, they're all over the map. Such "incoherent" light is like a crowd milling in Times Square: There's plenty of energy, but it's useless for practical purposes because it's chaotic and going every which way.

Enter the laser, a device for turning a mob into a fiercely disciplined drill team. Its name is short for light amplification by stimulated emission of radiation. Lasers produce "coherent" light in which all waves are exactly the same frequency, are precisely in phase, and are headed in the same direction. Certain crystals, some gases, and a few liquids give off perfectly coherent photons if you stimulate them in just the right way, by electric current or by another light source. In such situations, the atoms of lasing materials enter an odd, semi-stable, excited state that can persist several thousandths of a second—an eternity on the atomic scale. When an electron in one of those atoms gets energized by an incoming photon from a neighboring atom, it is prompted to emit a photon identical in every way to the one that triggered it. The two go flying off to hit other atoms, quickly producing four more identical photons, then eight, and so forth. Like a chain reaction, the population of coherent photons is amplified.

In surgery, laser beams can serve as scalpels, instantly cauterizing incisions and sealing broken blood vessels as they go. Dental researchers are testing silent laser "drills" that vaporize cavities instead of grinding at them. ∎

MIRROR

FLASH TUBE

RUBY ROD

Light that slices metal ▶
Laser light can be released in pulses or in a continuous beam. Either way, it may be powerful enough to cut steel, as in the case of this circular saw blade. The intensity of the light beam depends on the type of material and the design.

▼ Inside a laser
A ruby rod or other lasing material is surrounded by a stimulating light source, here shown as a spiral "flash" tube. Radiation emitted by the spiral flash tube excites the lasing material to give off coherent photons. Mirrors at each end of the laser chamber intensify this effect, causing the selected light to resonate back and forth in a concentrated path.

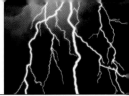

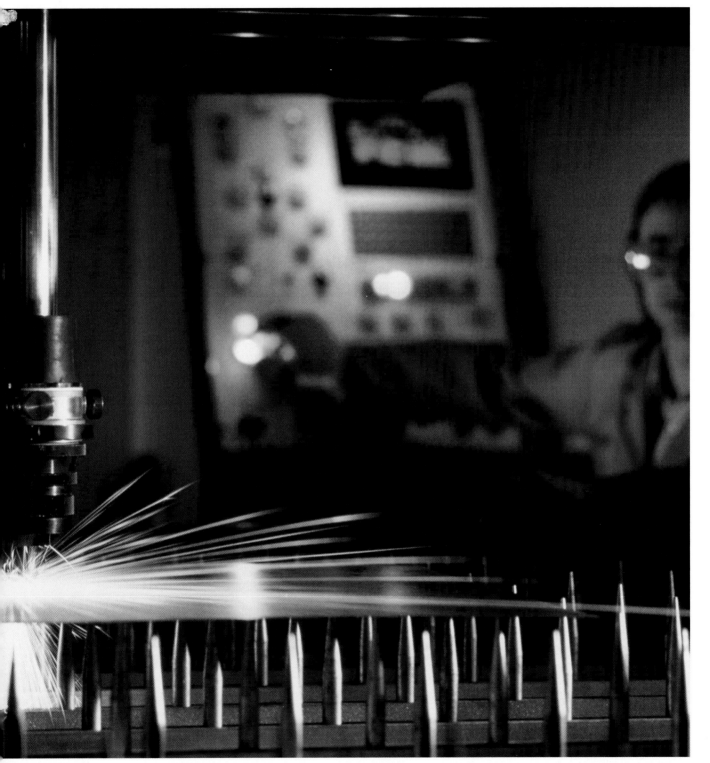

SEE ALSO

It's all downhill
from here
10-11

Making waves
56-59

Resonance and
interference
62-65

Magnetism
and our lives
84-85

Now you
see it...
100-101

Making light
speak
118-121

The photo-
electric effect
124-125

Making light speak

Although we often associate lasers with high energy, most are low-strength varieties, whose main advantages are accuracy and predictability.

f you've ever driven through thick fog on a country road at night, you know how fast the beams of ordinary incoherent photons from your headlights spread out and diffuse in the air. Coherent laser light, however, has its thing together. It stays so tight that scientists can shoot a thin beam at the moon—aiming precisely at a mirror left there by Apollo astronauts—and then measure the time it takes for the light to return, thus enabling them to determine exactly how the Earth-to-moon distance is changing over time. The beam they use scarcely spreads at all in the course of completing its 500,000-mile round-trip; and because it carries only one specific frequency, its reflected signal is easy to detect through the optical "noise" in the atmosphere.

But lasers find their way into much less exotic applications. Engineers who dig tunnels underground, and thus can't use the surveying techniques that work on the surface, aim their boring equipment with laser beams. If you've had a ceiling built or replaced in your home recently, the workers probably used a laser beacon to shine a thin reference beam onto the walls. And, of course, every CD or DVD player or read-write drive uses laser light.

One of the most exciting uses is in computing and communication technology. Because light waves are so small, and because it is relatively easily to get several wavelengths to share the same space, they're a natural choice to replace electrical signals in a new generation of optical computers. And although light travels at only about two-thirds of its maximum (vacuum) speed in glass, that's still drastically faster than comparatively sluggish electrons. ■

Wires for light ▶
One of the most important technological breakthroughs of the 20th century was the creation of efficient glass fibers to carry light signals. They're actually coaxial units, with one kind of glass wrapped around another like the insulation on a wire. This produces nearly total internal reflection of the light waves. Even then, the signals weaken over long distances, so they have to be boosted at intervals. Nowadays, each individual fiber strand in a bundle can carry billions of signals per second.

Checking out the code

Supermarkets and other retail outlets use laser scanners to read bar codes on products—both to speed up checkout lanes and to take inventory. Such beams usually originate from low-energy gallium arsenide lasers that give off infrared radiation. The system works because the laser beam stays tight even as it bounces off the bar code; since the detector reads only one frequency, it is unaffected by fluctuating light conditions in the store.

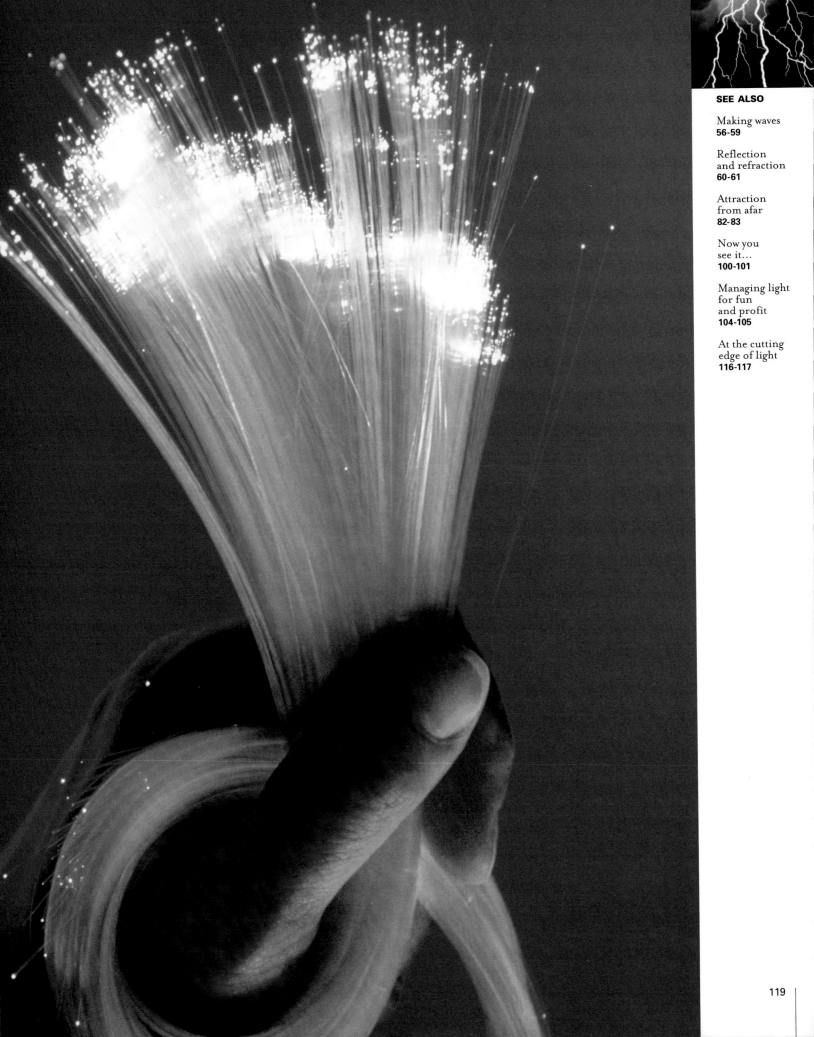

119

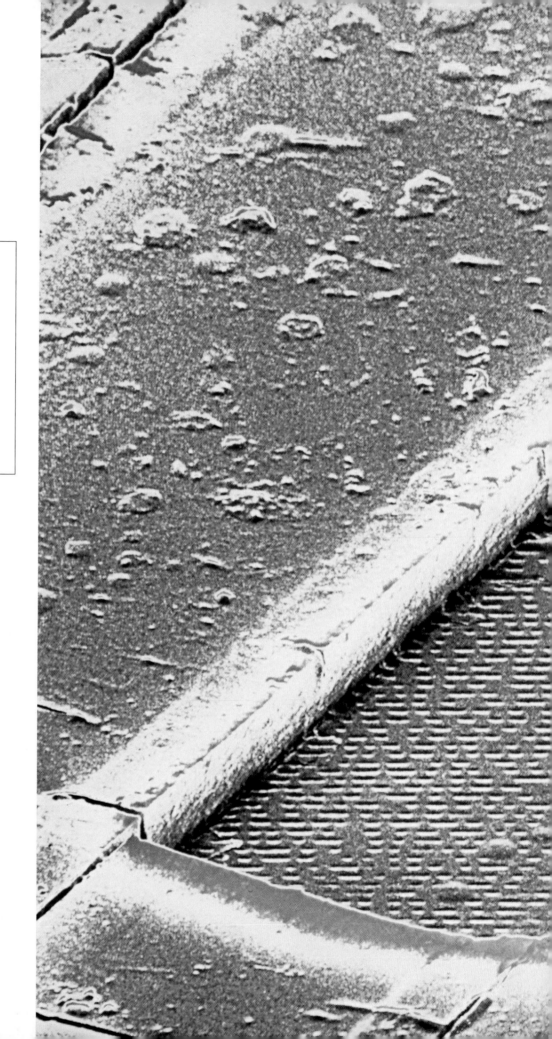

Lasers and compact disks

Essentially, a digital compact disk is a plastic plate that has been etched with a few billion microscopic pits. Flat areas reflect your CD player's scanning laser light back to a detector; the pits scatter the beam. Thus is created the on-off sequencing necessary for binary data. The plate is coated with very shiny aluminum and then covered with a transparent protective layer—part of which has been cut away in this image to reveal the inner structure.

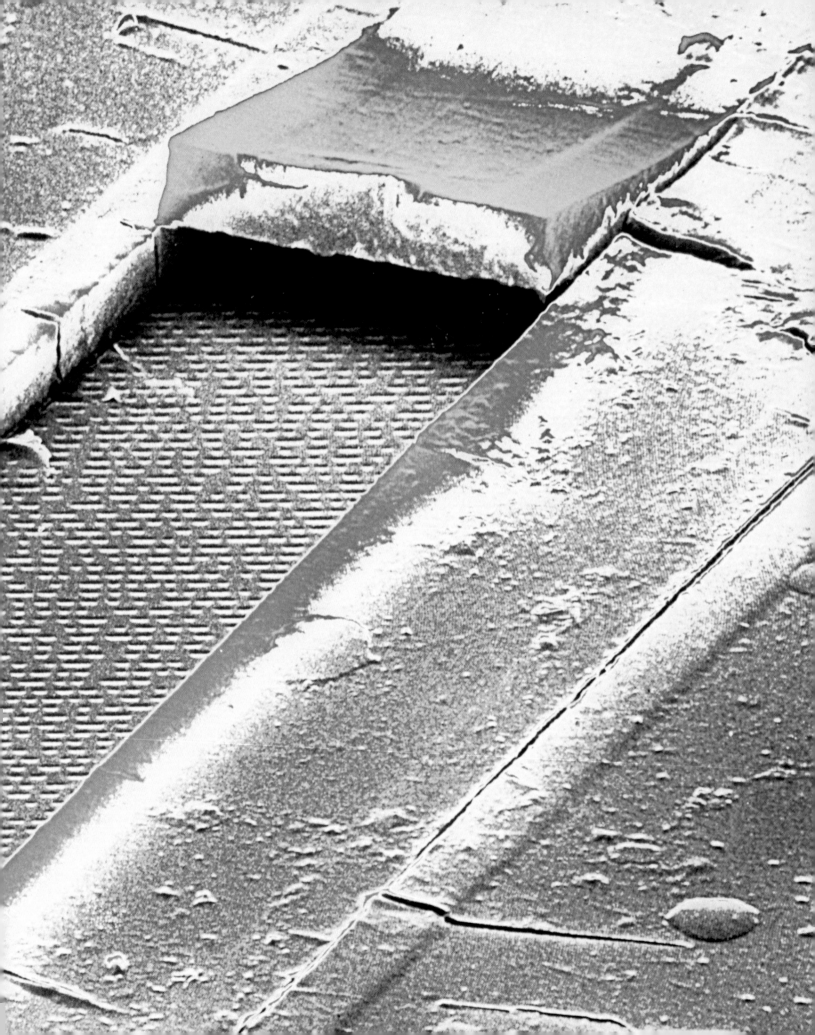

In our own image

Medical imaging now relies on multiple technologies beyond x-rays,
each method using a particular part of the electromagnetic spectrum.

When it comes to seeing inside your own body, the hard parts are really the easiest: Just take an x-ray. To get images of soft tissues, physicians look to other wavelengths and techniques such as magnetic resonance imaging (MRI) and positron emission tomography (PET) scans. Any charged atomic particle has a magnetic field that results from its spin. Ordinarily, those spins point randomly, in all directions. But crawl inside the powerful magnetic field of an MRI machine (far right) and the proton spins of hydrogen atoms—which are everywhere in your body, but at different concentrations in different tissues—tend to line up with that field. The machine then zaps you with radio waves of about 40 million hertz, precisely the resonant frequency needed to switch the magnetic polarity of your protons. When the waves are turned off, those atoms flip back—momentarily becoming tiny radio transmitters. Detectors in the MRI machine record those signals and use that information to compose a soft-tissue portrait. ■

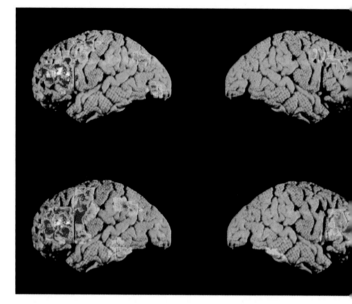

▲ Getting acquainted with PETS

Patients undergoing PET scans receive small amounts of a radioactive tracer. As each radioactive atom decays, it emits a particle called a positron that is instantly annihilated by an ordinary electron, producing two gamma rays that fly off in precisely opposite directions. A detector monitors where the rays originated. Tissues that absorb more tracer—such as active brain cells—produce more gamma rays. Thus PET scans show which parts of the brain are most active during a particular kind of mental activity or task.

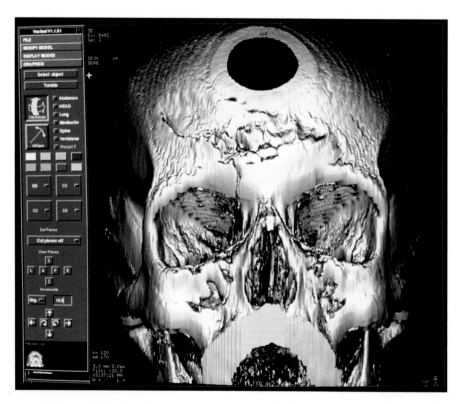

◄ Taking a CAT snap

Modern computers create three-dimensional views of tissues by taking multiple x-rays, each at a slightly different angle. Software programs then reassemble and superimpose the data into a series of images so that interior structures can be viewed at various angles and depths. That's computerized axial tomography, aka CAT or CT.

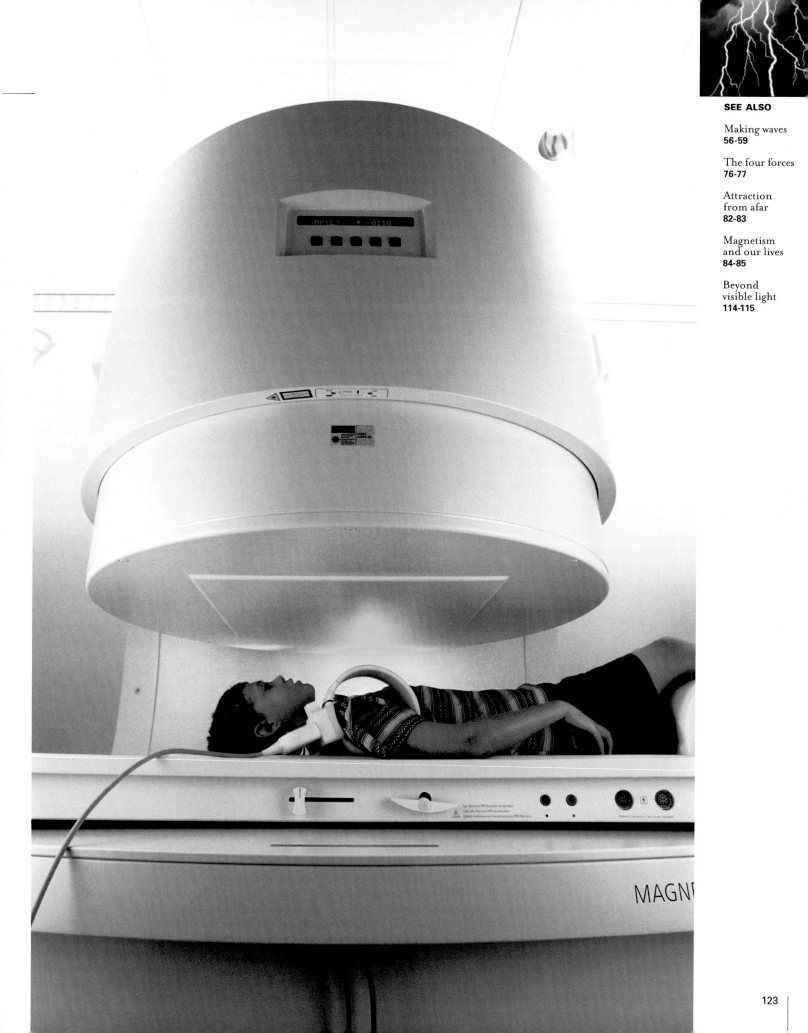

The photoelectric effect

For centuries, physicists argued whether light was a pure wave or a stream of discrete particles. They now believe that it's actually a little of both.

A lot of folks assume that Albert Einstein (1879-1955) must have received the Nobel Prize for his theory of relativity or for penning the world's most famous equation, $E=mc^2$. But in fact, he won it for explaining the photoelectric effect, a much more practical phenomenon that's at work in your camcorder or digital camera.

Earlier, scientists noticed that when light rays above a threshold frequency strike certain metals, enough electrons are ejected to create a measurable current. Einstein made the then-outrageous assertion that this could be understood by thinking of light as individual particles, or photons, hitting each electron and knocking it out of the metal. He added that each photon's energy was due to the frequency of its light, not the intensity. He was proved right, and got the Nobel. We got the electronics.

When coupled to switches, photoelectric devices can be used to open doors automatically or trigger alarms. A transmitter beam is aimed at a receiver some distance away; as long as light hits the receiver and throws off enough electrons to provide a current, nothing happens. But if the beam is interrupted by something opaque, the current disappears, which trips a switch and starts the action. ∎

Capturing motion and sound ▶ *is a snap with a video camera. Its imaging heart is a charge-coupled device, or CCD, whose silicon "film" makes use of the photoelectric effect. After being focused by the camcorder's lens, the image strikes a CCD, which is made up of as many as several million tiny light sensors. Photons striking each tiny pixel on the silicon grid dislodge electrons, and the pattern of light and dark is stored electronically on a magnetic tape or memory chip.*

▼ Einstein on Mars *Because the photoelectric effect results in an electronic signal (as opposed to the film in a camera), video cameras are ideal for instantaneous transmission. That's how NASA broadcast the exploits of its robot Mars rover in 1997, using a camera on the lander.*

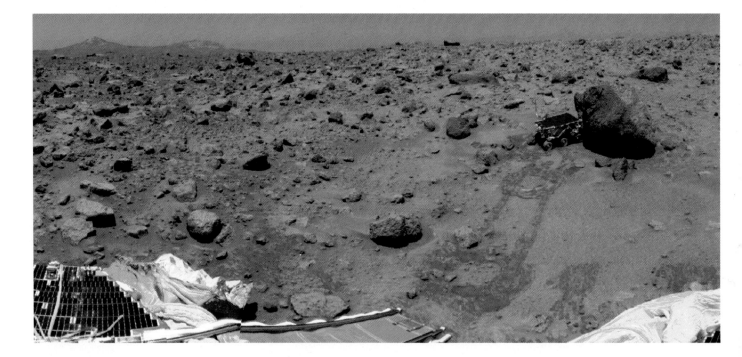

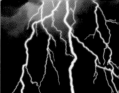

SEE ALSO

Making waves
56-59

The four forces
76-77

Now you
see it…
100-101

Managing light
for fun
and profit
104-105

At the cutting
edge of light
116-117

Making
light speak
118-121

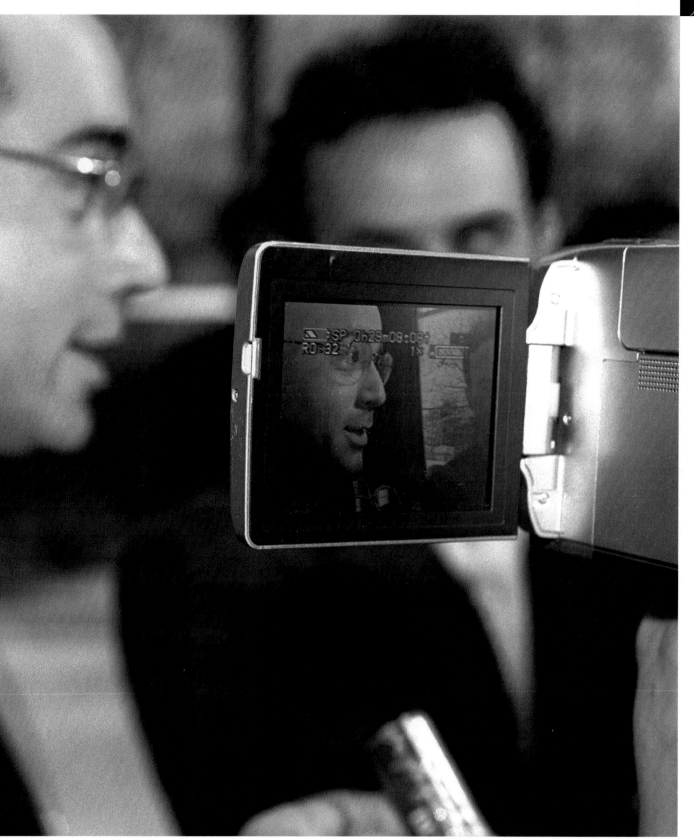

Inner sanctum

Deep within some atoms, things are just falling apart. That's what radioactivity is, and it shows that even nature's strong force has limits.

Radioactivity occurs because a few elements are just too big to be stable. The strong force—which works only at very close range to bind protons and neutrons in the atomic nucleus—can start to lose its grip when the nuclear count goes beyond 82 protons, because at that point the natural electrical repulsion of so many positively charged protons really begins to add up. So certain kinds of atoms—especially big ones—break down, throwing something off in the process. Often, their emissions can set off chain reactions.

There are three main kinds of emissions. One is a clump of two neutrons and two protons called an alpha particle. Radium emits alphas, which can be stopped effectively by your shirt sleeve or a sheet of typing paper. But if alpha emitters get in your lungs, where there is no protective barrier, the particle bombardment can cause cancer. That's why you may have had your home tested for the presence of radon, a radioactive gas produced naturally when radium decays. A second type of emission is an energized electron, also called a beta particle. A thin sheet of aluminum can keep betas away. The third form is the gamma ray—at the short-wavelength, high-frequency end of the electromagnetic spectrum. To hold them off, you need a couple of inches of lead.

Probably your only radioactive device is a smoke detector. It uses a manmade radioactive element, americium, to generate alpha and beta particles, ionizing the air and allowing a current to flow between two electrodes. When smoke interrupts the current flow, a detector senses it and sounds the alarm. ∎

Hot stuff ▶
Nuclear reactors do not create carbon dioxide or other greenhouse gases. But they do produce two waste products. One is excess heat, which is absorbed by a closed system of circulating water and is shed into the air via cooling towers. The other is spent fuel, in the form of rods. Finding a way to dispose of those rods is a major political and engineering problem.

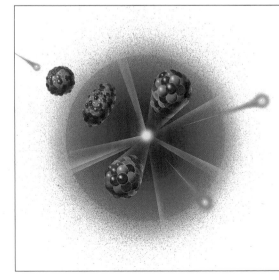

The power of fission

More than 50 years ago, nuclear fission incinerated Hiroshima and Nagasaki. Now it produces about one-sixth of the world's electric power, at over 400 nuclear plants in 30 different nations. In most reactors, atoms of uranium-235 split into two roughly equal halves (left), tossing off two or three neutrons. Each neutron strikes another uranium atom, which in turn splits and gives off more neutrons, in an accelerating chain reaction. Usually, the heat of this reaction is used to boil water, and the steam turns turbines to generate electricity.

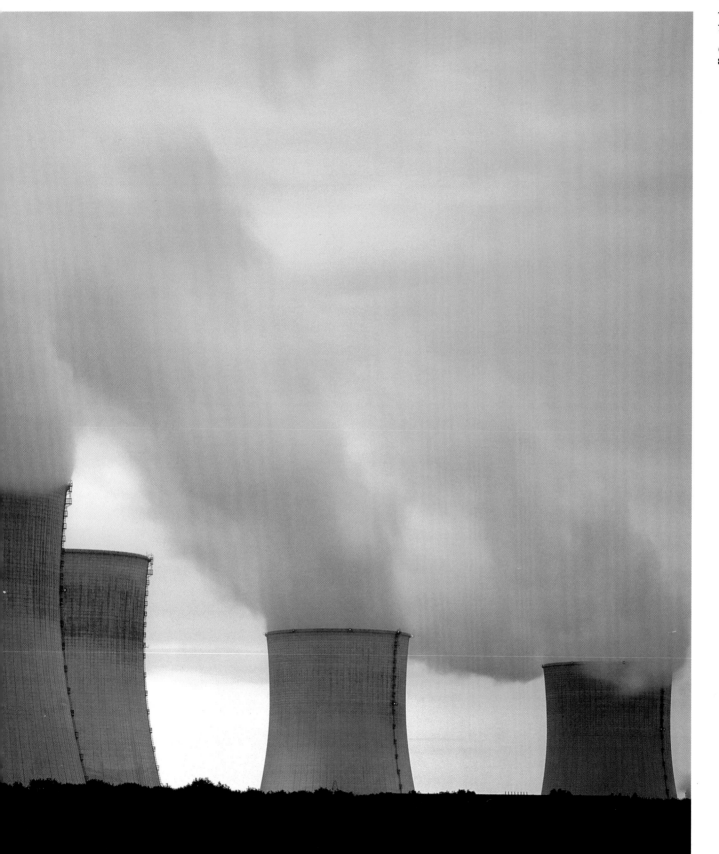
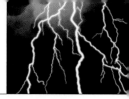

SEE ALSO

The four forces
76-77

Getting charged
86-87

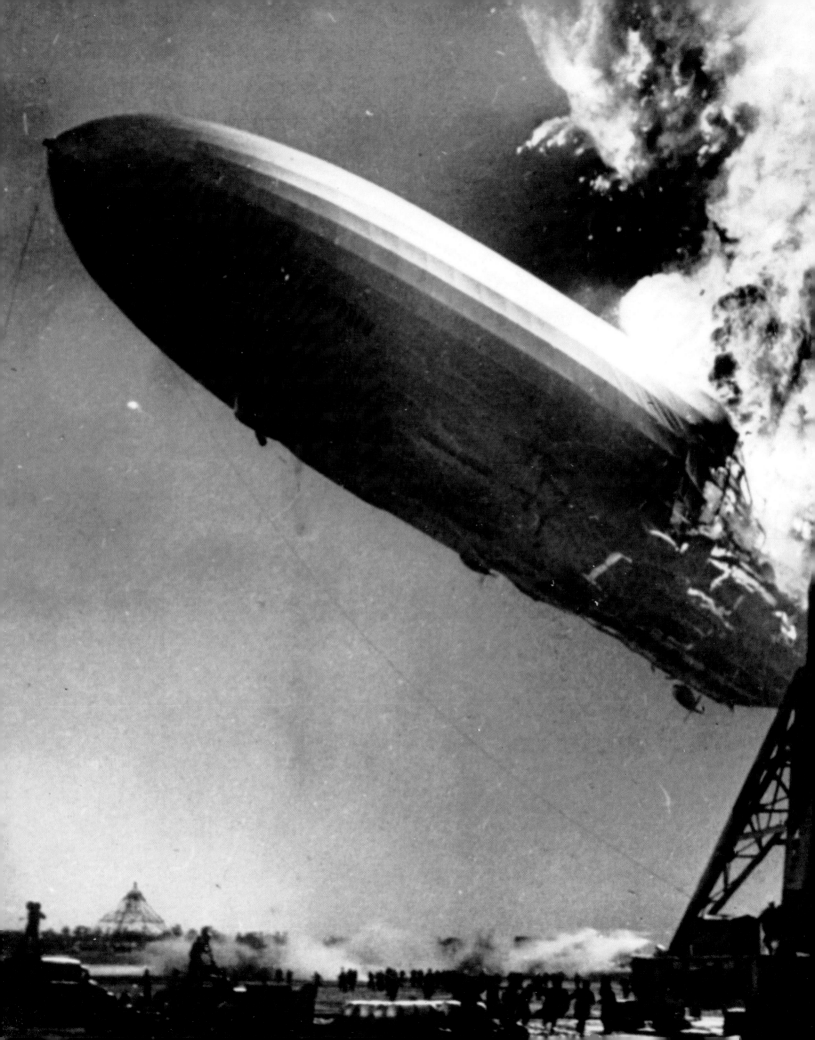

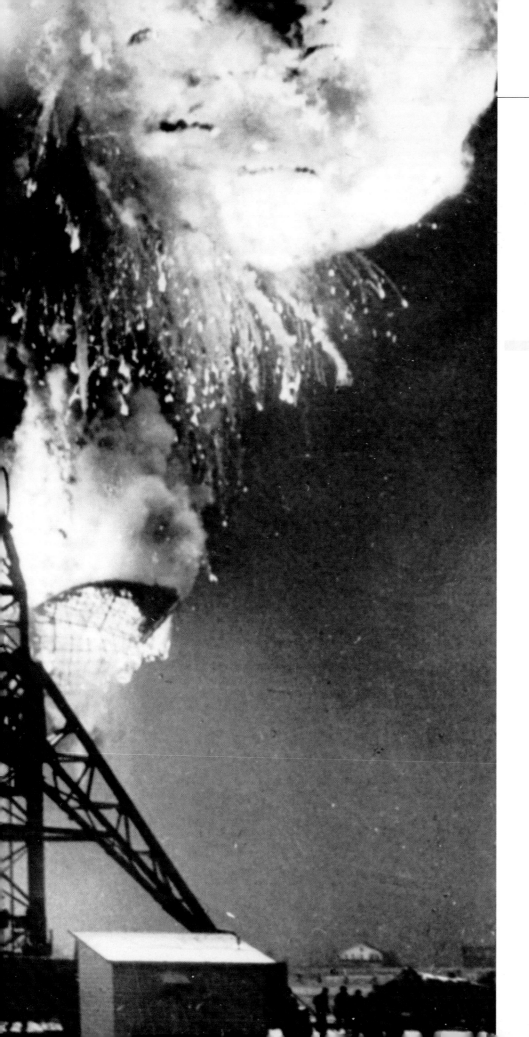

The
Right
Stuff

he material world, which seems so solid and stable, is in fact incessantly churning with change. Often the processes are slower than rust. Sometimes they're nearly instantaneous. In the 1937 explosion that destroyed the dirigible *Hindenburg,* hydrogen gas leaking from the airship combined violently with oxygen in the air, turning the sky to flame. But if that gas had been helium—an element which differs only slightly from hydrogen yet is still seven times lighter than air—the disaster could not have occurred. For unlike hydrogen, helium cannot burn. By understanding the nature of elements, we utilize and control chemical reactions to make the thousands of substances we use every day.

Elements: personality kids

Just over 90 different elements occur naturally; nearly everything we see and touch is made up of combinations of the three dozen most common.

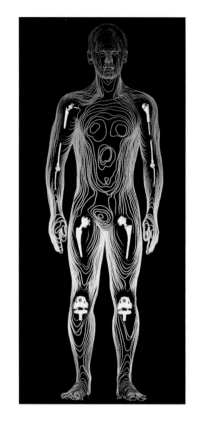

Each element is made up of atoms that are unique for that substance: They have a certain number of protons—and a matching number of electrons to keep them electrically neutral. Yet some elements behave in surprisingly similar ways. Just as high school students tend to fall into distinctive categories—clowns, socialites, jocks, nerds, delinquents, or loners—chemical elements display definite personalities, from highly reactive to easygoing to downright inert. It's all a function of how their electrons are arranged. Chemistry is the science of how atoms and molecules share, shuffle, or exchange electrons to make or break bonds. Electrons are arrayed around the nucleus in ever wider orbital levels or "shells," each of which can hold a fixed number of electrons: a maximum of 2 in the first level, 8 in the next, on up to 32 in large atoms. In general, the closer an atom is to having a full outer shell, the more likely it is to shed or grab electrons—and the more reactive it is. Those with complete shells, such as helium atoms, almost never interact with their neighbors. Those with only one electron to gain (say, chlorine) and only one to share (for example, sodium) are natural partners. They come together readily in a frenzy of complementary needs—resulting in compounds such as sodium chloride, otherwise known as table salt. ■

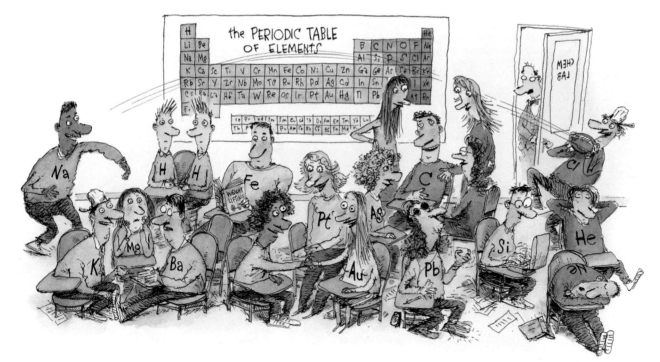

◄ **It takes all kinds**
In essence a personality chart, the periodic table groups elements according to the configuration of electrons in their outermost shells, which basically determines their behaviors. Similar characteristics tend to recur at regular intervals (the "periods" in the table), and elements with those particular characteristics are arranged in vertical columns.

◄ **Hardware**

Some elements don't react with biological systems—titanium, for example. It's also lightweight and strong, and therefore ideally suited to replacement parts for bones and joints.

Foiled again... and again ►

Aluminum, with hundreds of uses from kitchen foil to aircraft bodies, ranks as the third most commonly found element in the Earth's crust (after oxygen and silicon), making up a whopping 8 percent of the crust by weight. Next in order of abundance are a host of reactive metals: iron, magnesium, calcium, potassium, and sodium. All other elements account for less than one percent.

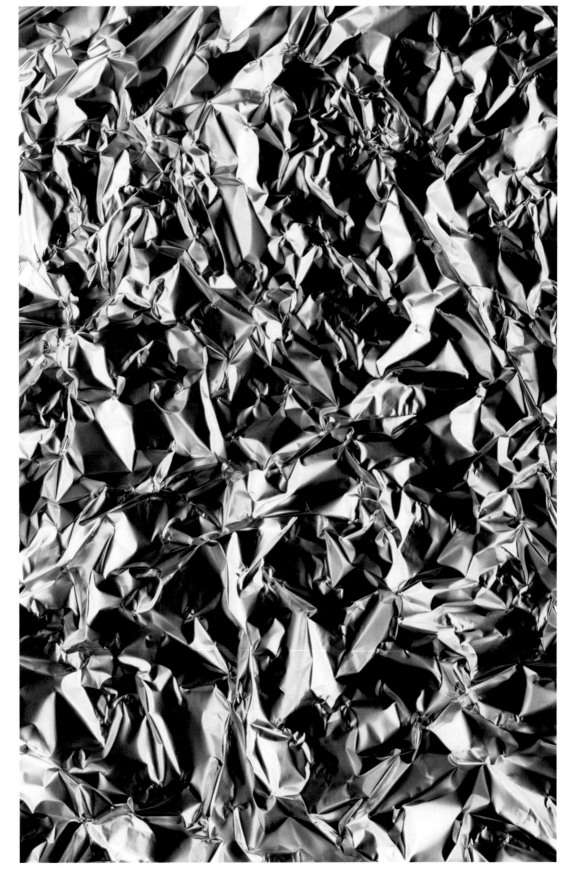

Getting together

Ripening human relationships "have the right chemistry," we say, and so it is when elements get together. Different partners combine in different ways.

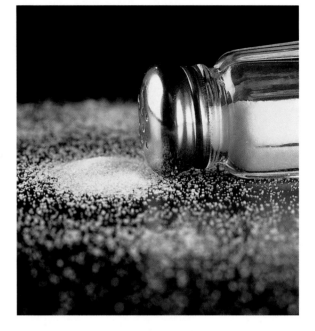

◄ **Strange offspring**
Sodium is a butter-soft, gray metal so caustic it can burn your hands. Chlorine is a greenish, poisonous gas. But combine them into sodium chloride, or salt, and the result is a clear crystal compound that is absolutely essential to the chemistry of human life.

n the case of table salt, the attraction is pure electricity: Sodium loses an electron and acquires a positive charge, while chlorine gains an electron and becomes negative. The opposite charges slam together in what is called an ionic bond. Less impulsive elements often prefer sharing electrons in order to satisfy mutual needs; they form what are called covalent bonds. Oxygen, for example, has six electrons in its outermost shell and needs two more to fill it. Hydrogen has one and needs one more. Get two hydrogen atoms to share their electrons covalently with a single oxygen, and all three feel fulfilled.

Like atoms, molecules find all sorts of ways to combine. Water and salts form solutions—brews whose components may vary in ratio but can't be separated by filtration or other physical means. Some molecules merely tolerate each other, joining up in mixtures that can differ greatly in stability. The alcohol in beer stays fairly well dissolved, for example. But oils and water-based solutions are notorious non-mixers, as you find with bottles of salad dressing or rain on a greasy driveway. Milk is an oil-and-water combo known as a colloid; its tiny fat droplets are so finely dispersed that they hang suspended in the water. ■

Virtual hermits ►
Six elements are virtual hermits, known unflatteringly as inert gases. They have full outer shells. But hermits have their uses, especially when reactions are unwanted. Helium resists fire, and argon safely fills incandescent lightbulbs. Neon, krypton, and argon, confined in glass tubes and excited by electric current, help provide a city's multicolor dazzle.

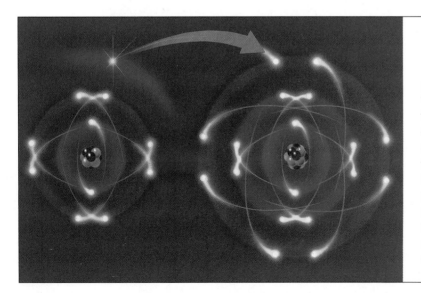

The electric attraction of ionic bonds

Some chemical bonds are called ionic because they arise from the attraction that occurs between two ions of opposite charge. Take our fidgety friends at the far ends of the periodic table, the halogens and the alkali metals. An atom of sodium (left) loses its outermost electron, becoming a positively charged ion because its 11 protons are now surrounded by only 10 electrons. When a chlorine atom gains sodium's outer electron, it becomes a negative ion, with 18 electrons racing around 17 protons.

Odd couples

What you might logically expect is not always what you get whenever atoms and molecules choose to mix it up.

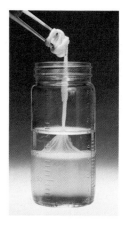

▲ **Hanging together**
Chemistry took a giant leap in the late 1930s when scientists learned to link carbon molecules together in long chains called polymers. One truly useful result: nylon.

Some of nature's most common characters pull off some oddball surprises. For example, hydrogen is the lightest gas known; oxygen is the fourth lightest. So why, when you put them together, do you get water—a fairly heavy liquid? Or try this: Carbon is a solid. Combine it with oxygen, which has even more protons and neutrons, thus a higher atomic weight—and you get a heavier solid, right? Wrong. You get buoyant carbon dioxide gas.

The reason for these wildly counterintuitive outcomes is that the state a compound takes—solid, liquid, or gas—can depend more on the way its molecules stick to each other than on the actual mass of the constituents. Carbon is normally a solid not because it's inherently ponderous—it's the sixth lightest element—but because carbon atoms hang on to each other with strong covalent bonds and form large agglomerations. Water is a liquid for a similar reason: Its molecules cling to each other through a kind of electrostatic force called a hydrogen bond.

Of course, if there's enough thermal energy around, the normal bond strength of a substance may be exceeded. Raise the temperature of water high enough and the hydrogen bonds will weaken, allowing the molecules to physically separate and become

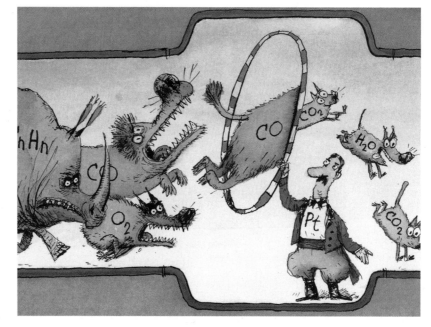

◄ **Magicians who never change**
Catalysts are substances that cause chemical reactions without themselves changing in the process. The catalytic converter in your car's exhaust system uses palladium or platinum, which prompts incompletely burned fuel to combine with oxygen, producing less harmful compounds.

gaseous steam. When your barbecue grill is blazing, even some of the carbon atoms tightly bound within a piece of hamburger will shake out and combine with air to make carbon dioxide. Covalent bonds are about ten times as strong as hydrogen bonds, and so higher temperatures are usually required to break them. Lighting the fuse on a string of firecrackers provides enough extra energy for the carbon, potassium, nitrogen, and sulfur in ordinary gunpowder to combine with oxygen in an explosive rearrangement. ■

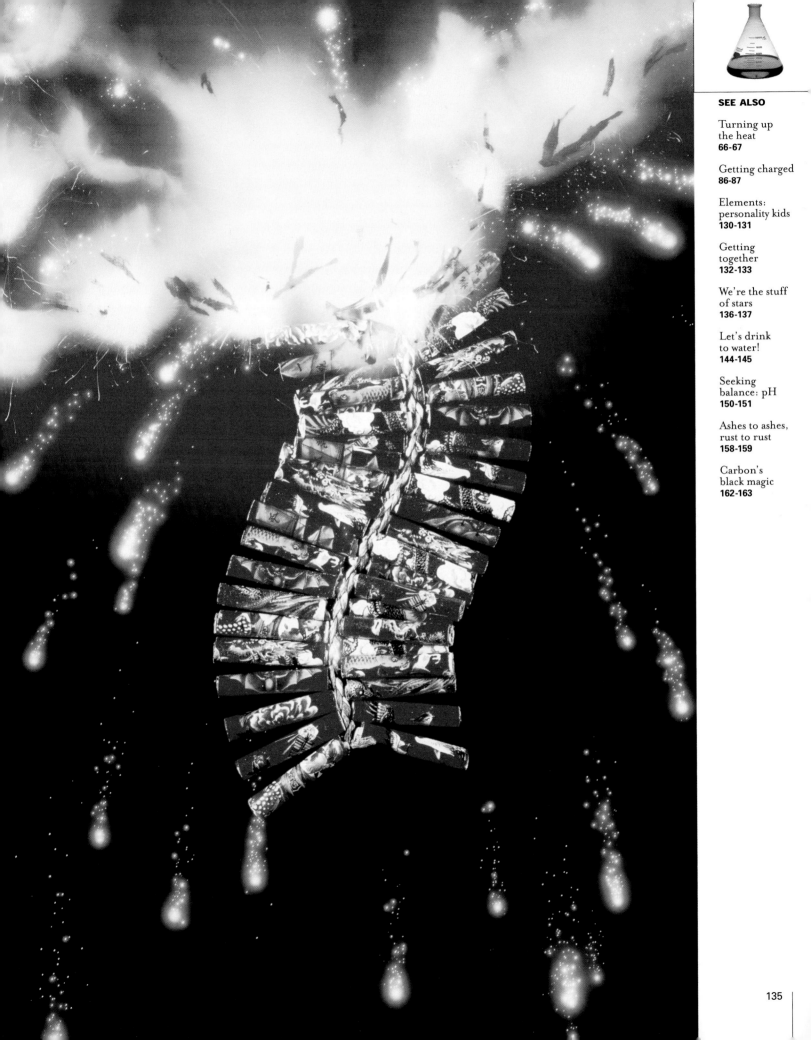

We're the stuff of stars

Get real. Stars aren't just in the sky—or just in Hollywood. Every one of us is more stellar than we may realize.

Stars are our ultimate ancestors. That's because they created the very elements of which we and our surroundings are made. More than 99 percent of the immensely scattered matter of the cosmos consists of the three lightest elements: hydrogen, helium, and lithium. They formed when the whole show got started some 14 billion years ago, in the natal cataclysm known as the Big Bang. Most other elements are manufactured in stars through nuclear fusion, whereby small atoms are merged into larger ones. Fusion also converts tiny amounts of mass into energy, which is why stars are luminous. When the big ones burn out, their mass collapses and then explodes in a spectacularly violent event called a supernova. When that occurs, the star's contents are strewn across neighboring space.

There was a lot of that debris in the gas ball from which our solar system condensed some five billion years ago. Most of the gas collected in the center, eventually forming the sun. Solar radiation heated up the innermost planets until they could no longer hold light elements such as hydrogen and helium. Jupiter, Neptune, and other gas giants were both distant and massive enough to retain their gases. ■

The Big Bang ▶

Scientists generally agree that our universe (still expanding fast in all directions) arose from a single point. From that origin, energy and matter erupted, expanded, and gradually cooled into the cosmos we see today.

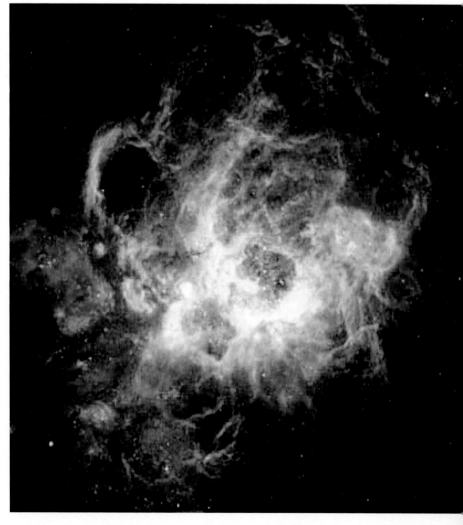

Star nursery

The same cosmic events that led to our solar system's birth five billion years ago appear to be taking place in the NGC 604 nebula (right), an immense cloud of gas, dust, and stars located in the spiral galaxy M33. Over two hundred stars dwell at the heart of this glowing nebula, most only about three million years old—mere toddlers by stellar standards. These stars are the hottest and brightest that could be found in any galaxy—about 100,000 times brighter than our sun. NGC 604's elements include not only hydrogen and helium but also carbon, nitrogen, and oxygen—plus molecules of water and carbon monoxide.

▼ Birth of the sun

Vast disparities characterize the distribution of elements between the sun and its third planet, Earth. Hydrogen and helium, so dominant in the sun, were largely driven from the young Earth by solar heat, leaving a large supply of the heavier elements that now support life.

▼ Complex Earth

As the Earth (and other planets) formed, it was bombarded by smaller blobs called planetesimals, millions of which rained down on the airless, infant Earth. The heat from those impacts—together with the planet's own thermal energy—in time raised temperatures to the point at which iron melted and began sinking toward the planet's core. Since iron makes up about a third of Earth's mass, this was not a trivial event.

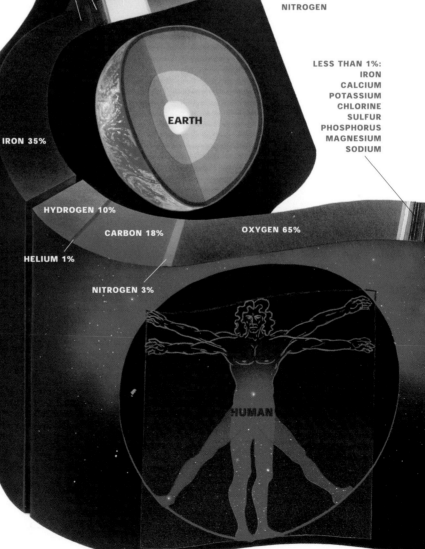

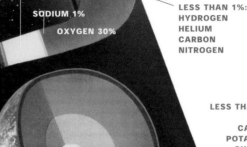

PHOSPHORUS 1%
SULFUR 2%
CHLORINE 1%
POTASSIUM 1%
CALCIUM 1%

MAGNESIUM 13%

SODIUM 1%

OXYGEN 30%

LESS THAN 1%:
HYDROGEN
HELIUM
CARBON
NITROGEN

HYDROGEN 75%

SUN

HELIUM 25%

IRON 35%

EARTH

LESS THAN 1%:
IRON
CALCIUM
POTASSIUM
CHLORINE
SULFUR
PHOSPHORUS
MAGNESIUM
SODIUM

HYDROGEN 10%

CARBON 18%

OXYGEN 65%

HELIUM 1%

NITROGEN 3%

HUMAN

Elements of man ▶

Computations show that if the Earth had been 7 or 8 percent closer to the sun, it would have been too warm for much water to have condensed into liquid. If it had been significantly farther away, it would—like Mars— have remained perpetually frozen. Oceans could not have formed, and life as we know it would never have arisen. By equally good fortune, the planet's surface contained abundant supplies of the most important elements in our bodies. Earth also had a moon that provided tides, which may have helped the planet's first life-forms arise in tidal pools.

Skeletons of substance

*Some things are crystalline, some are amorphous. The type of structure—
and its basic inner strength—depend on just how the atoms bond.*

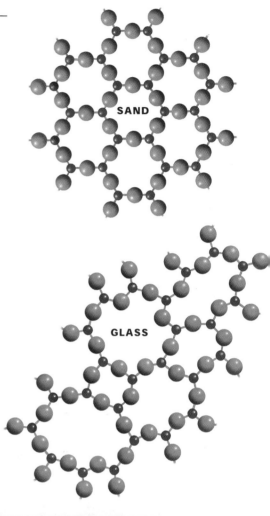

SAND

GLASS

Hair combed, clothes pressed, shoes shined—the neatniks in chemistry class
find ready counterparts in crystalline solids, which consist of tidy lattices of
interlocked atoms in regular, geometric arrangement. Then there are the
class slobs—hair all over, jeans rumpled, tennies untied—the amorphous or
non-crystalline materials, solid but structurally far less prim. Most solids are
crystals; such structure is what makes metals strong, rocks hard, and quartz
faceted. Amorphous solids include things like glass, rubber, and porcelain.

 Crystal solids divide into four types, determined by the sort of bond that
holds them together. We've already seen how table salt forms ionic bonds. So
do things like calcium carbonate, the principal ingredient in limestone as
well as a likely component of your toothpaste. The second type, covalent crystals, form hard, durable structures such as diamond and quartz, a compound
of silicon and oxygen. Molecular crystals such as table sugar are third; they
usually make for soft materials, because their components are held together
by comparatively weak electrostatic or hydrogen bonds. The fourth kind, metallic crystals, often form cubic or hexagonal patterns that are very densely
packed, which gives metals their intrinsic heft. ∎

Amorphous cube

Knowing the non-crystalline structure of
glass helps engineers design high-tech
synthetics with a wide range of desired properties. One example is this glass ceramic used
to make the tiles that protect much of the
outer surface of the United States' space
shuttles from the ferocious heat generated by
air friction during re-entry. The tile is 93 percent air, and such a magnificent insulator that
even when heated to 2,200°F, it can be held
safely by the corners only seconds later. At the
other end of the conductive spectrum is diamond, which sheds heat rapidly for the same
reason that it is the hardest naturally occurring substance: Its crystal array is made from
repeating patterns of four carbon atoms, each
of which is bound tightly to its neighbors in a
pyramid shape by double covalent bonds.
This tetrahedral structure is very strong.

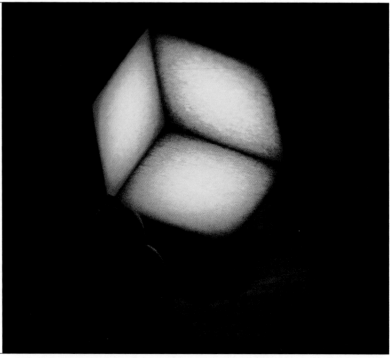

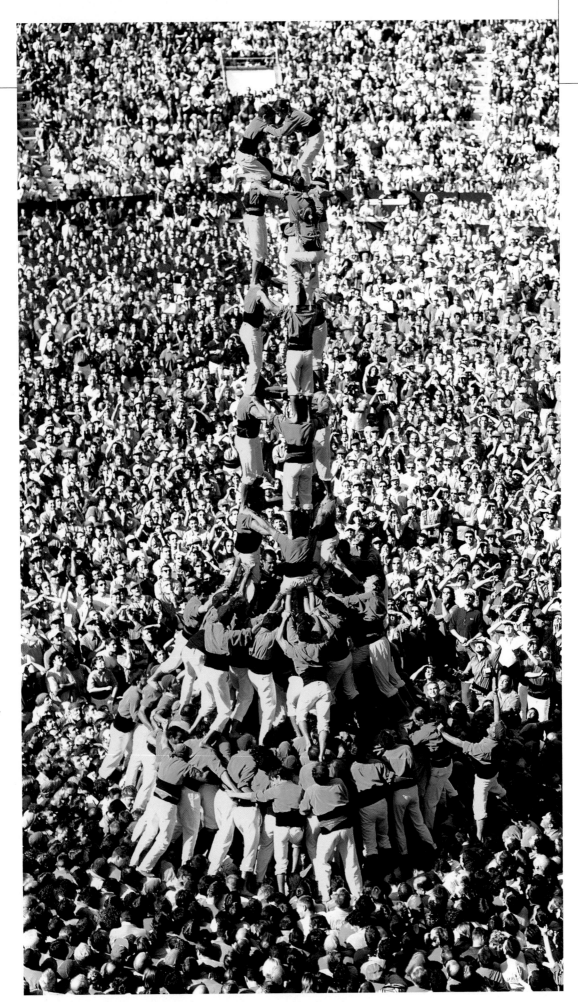

◄ **All in the cooking**
*These two atomic
skeletons are
chemically identical—
both silicon dioxide,
or SiO_2. But
structurally they are
very different. Beach
sand is crystalline;
glass is amorphous.
Think of glass as a
crystal that never got
to form because it
cooled too fast.*

Crystal power ▶
*Crystals derive their
structures (and their
strength) from the
size of the atoms that
combine to form
them, the strength
of the chemical
bonds among the
constituents, and the
pattern of the array
or "lattice" formed by
the constituents. All
have in common a
regular, repeating
pattern such as the
groupings of people
at the top of this
human pyramid.*

139

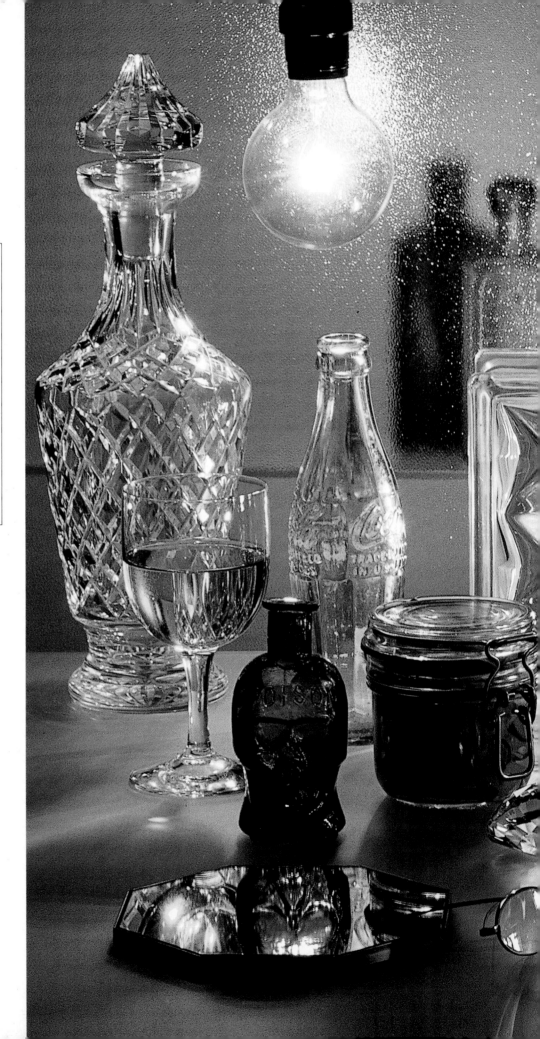

A transparent world

The characteristics of glass—like those of different steels—often stem from various additives rather than from its main component, silicon dioxide. Cobalt turns glass blue; chromium makes it orange. Gold and copper go into red "ruby" glass, and lead gives so-called "crystal" its higher index of refraction and sparkle. Boron makes a glass that expands and contracts less, thus is less likely to fracture during sudden temperature changes. Such "Pyrex" glass is ideal for the kitchen. Adding sodium and calcium yields a glass that melts more easily but can't be penetrated by liquids.

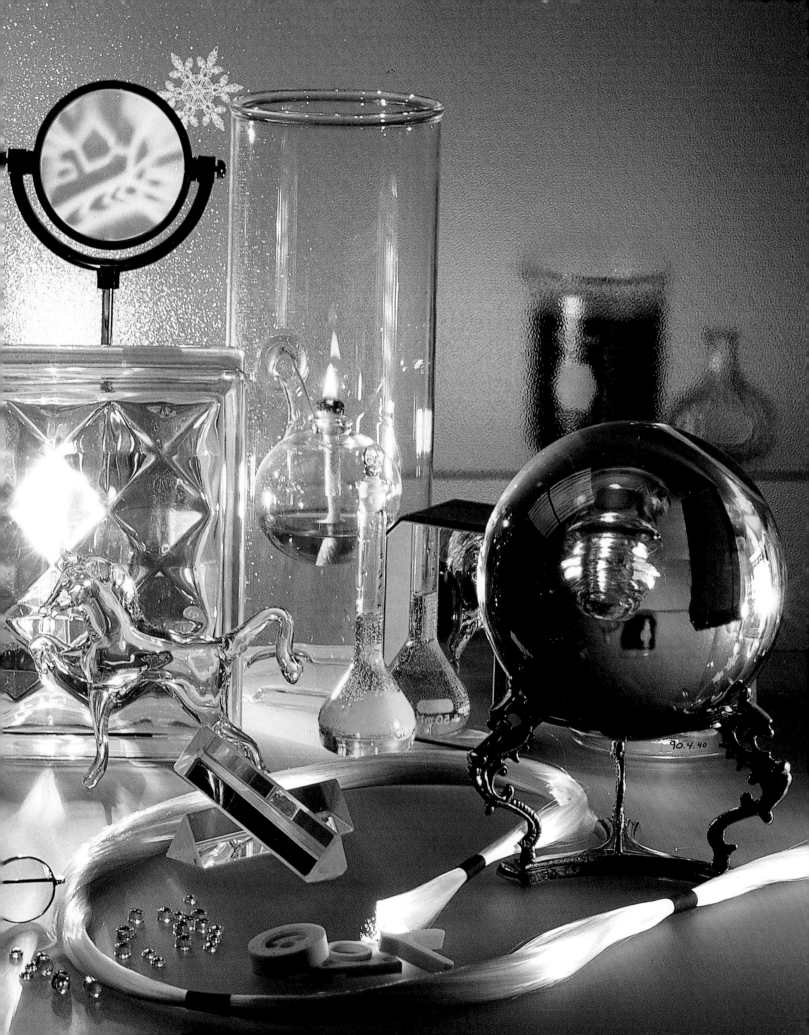

90.4.40

Marriages of convenience

*True, alloys lack the intimacy of metals wedded in chemical compounds.
But just the same, their relationships wear well and make for a better world.*

Combine two molten metals, stir them, and you can get a third, known as an alloy. Often its personality will be entirely different from that of either component. The atoms in alloys usually are not chemically joined, but exist in simple mixtures—like different liquids in mutual solution. Each component lends its own characteristic to the crystal lattice. Because metals make up three-fourths of all elements, there are plenty of possibilities.

Even some nonmetals can take part. Mixing iron with a nonmetal like carbon, for example, makes steel—an alloy that is much harder than iron. Civilization's progress is so intimately bound up with the ability to alloy metals that we describe an entire historical epoch as the Bronze Age, the period when our ancestors learned to mix copper and tin to produce a relatively strong material that resisted corrosion and was easily cast into tools and weapons.

Metals such as sodium and potassium willingly part with their outer-shell electrons. But some are notorious tightwads. Three familiar ones are copper, silver, and gold—whose electrons are strongly attracted to the positive nucleus, requiring a great deal of "ionization energy" to pry them away and make compounds. That's why gold and silver are usually found in a pure state in mines. This low reactivity makes alloys of gold, silver, and copper splendidly suited for use as dental crowns, coins, and electronic connections. Gold alloyed with nickel makes hard, highly corrosion-resistant "white gold." ∎

Bare bones ▶
Numerous alloys are used for bone, joint, and other surgical replacements, including alloys of titanium, cobalt, chromium, vanadium, and molybdenum. To ensure tight fit and high accuracy, hip implants can be placed with computer-assisted surgery.

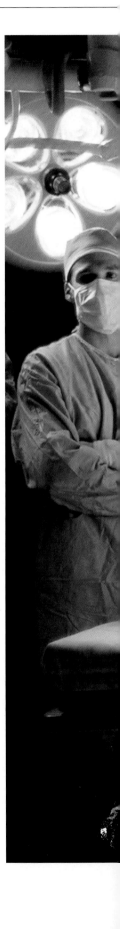

The many faces of steel

Mixing iron with carbon makes steel, which is both harder and more flexible than iron alone. But that's just the beginning. Carbon can be chemically bound to the iron as iron carbide, or it can settle into strands of graphite, depending on how fast it cools. High carbide content makes a very hard but brittle steel; high graphite content makes it more malleable. In addition to carbon, the stainless steel found in razor blades, flatware, kitchen implements, and surgical equipment also includes manganese, phosphorus, sulfur, silicon, nickel, and chromium. High-strength steel used in construction has all of those elements plus copper.

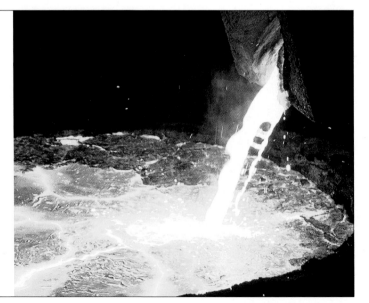

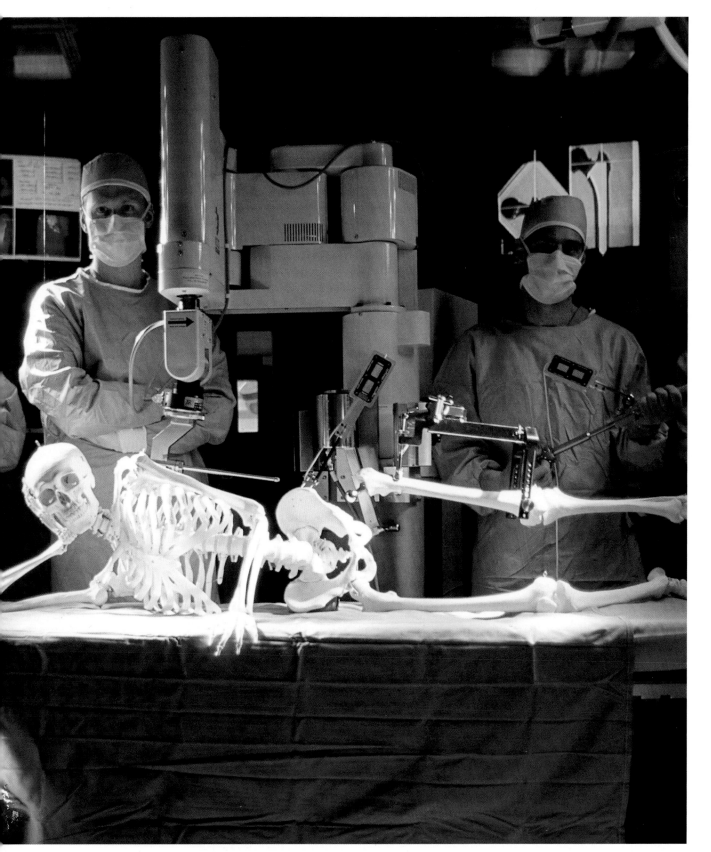

143

Let's drink to water!

Of all the ingredients essential to our existence, the one we probably take most for granted is in many ways the most wonderful.

We are water creatures, you and I. Water is our ancestral abode—after all, life began in the sea; the briny tang of blood may reflect that heritage. Water makes up about 60 percent of our body weight, and it serves as the medium of exchange for our most critical physiological functions. It's something we just can't be without. It's also one of very few substances that exist in all three states—solid, liquid, and gas—within the rather narrow temperature range that human beings tolerate, and it has an enormous ability to absorb heat.

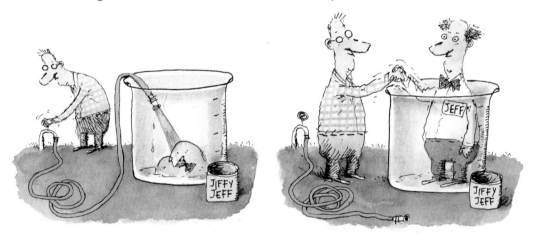

It takes nearly five times as much energy to raise the temperature of water by one degree Celsius as it does to heat aluminum the same amount. That's why pots and pans on your stove can get too hot to touch even though the water in them is merely warm. This high "specific heat" helps keep your body temperature stable and buffers drastic changes in the planet's air temperature. Unlike virtually all other compounds, water's solid form is less dense than its liquid form. Thus ice cubes and icebergs float—a matter of momentous importance to aquatic life, and to us.

One of the weirdest things about water is that it's a liquid at room temperature. The molecule itself is extremely light—about the same as a molecule of nitrogen or fluorine. So you might expect it to be a gas. But a water molecule's propensity to bind to neighboring water molecules makes it clump together into a liquid. The same property accounts for water's comparatively high freezing point.

Water is such a superb solvent for biological purposes that is is considered essential to life. Much of the excitement surrounding exploration on Mars comes from evidence that the Red Planet once had considerable water. ∎

Water world ►
Visit Yellowstone National Park in winter and you'll see water in its three natural states—solid snow and ice, liquid thermal springs, and gaseous mists. The actual molecules don't change much; they look like Mickey Mouse heads in all three states (inset), with a large oxygen atom flanked by two hydrogen "ears." This diagram shows the neat molecular lattice of ice above free-moving water molecules, with a transition zone of slush. Water's odd ability to be less dense as a solid than as a liquid reflects the fact that the hydrogen atoms shift slightly in angle as ice crystals form, forcing an increase in volume.

Hard waters

Plain water, containing nothing but H_2O, is remarkably rare in nature. Even so-called "pure" spring waters typically contain calcium, iron, or magnesium salts that derive from the soil or rocks through which the stream water runs. When the concentration of such salts is greater than one part in 10,000, the water is called "hard." The results can be seen as the mineral scale that often lines the insides of teapots or water heaters. The dissolved metals also interact with soap to form a curd-like scum—the dreaded bathtub ring—that is insoluble in water, and they reduce the soap's ability to make lather.

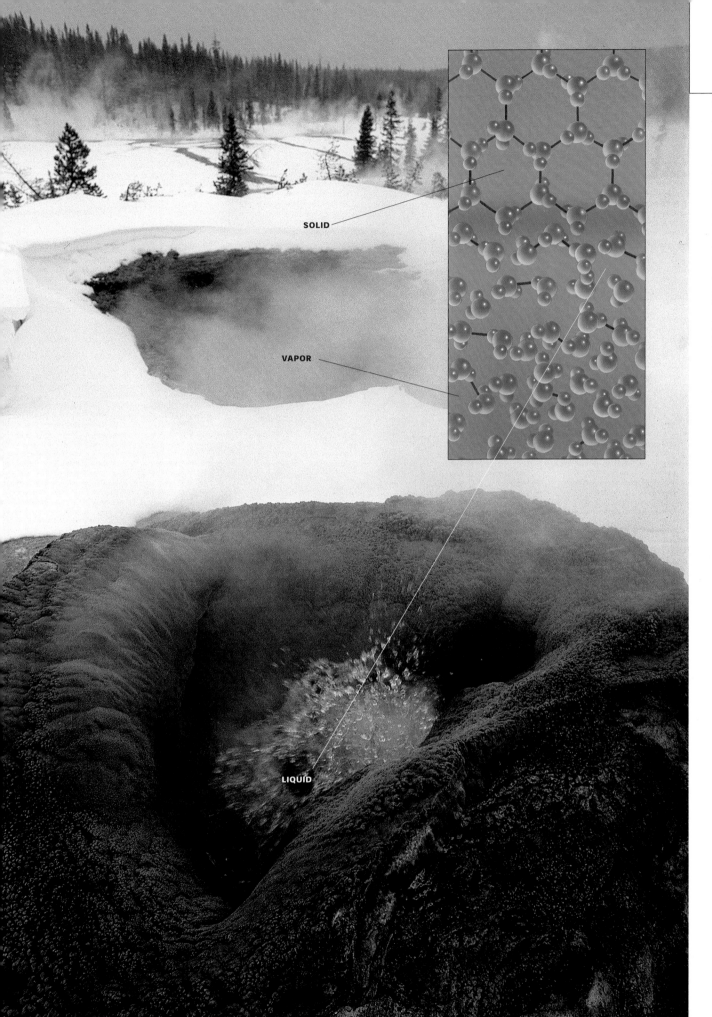

SOLID

VAPOR

LIQUID

Fresh water: not here

It sounds good, but "fresh" water basically doesn't exist; it's been around and around so many times that you could consider it the ultimate recycler.

Next time you leave the faucet running for minutes at a time while you brush your teeth or soap your hands, consider this: A mere one one-hundredth of one percent of all the water on the planet is readily available fresh water. About 97.4 percent of Earth's total water supply is in the oceans; some 2.6 percent is locked up in polar ice or in deep underground aquifers.

Of course, there's really no such thing as "fresh" water. Molecules of water pass through endless cycles—evaporating into the atmosphere, falling as rain, and running back to the sea with new loads of dissolved minerals—but there are no outside sources. The odd icy comet or meteorite hits our atmosphere now and then, but basically all our water has been here for about 4 billion years, ever since early volcanic eruptions and "outgassing" presumably liberated huge amounts of trapped water and volatile gases, creating the seas. ■

The water cycle

Each day, solar power evaporates a trillion tons of water from the planet's surface and pumps it into the atmosphere. Each day, the atmosphere surrenders the same amount of vapor, condensed as rain, sleet, snow, and hail. This endless renewal is the water cycle, and it represents the greatest physical force at work on the planet. Thus, while your drinking water is constantly recycled and never really fresh, it is continually purified by the cycle. Of all the water that evaporates, about 86 percent of it comes from the oceans, and the remaining 14 percent from land. But land masses get 22 percent of all precipitation that falls from the titanic distillery that we call the hydrologic cycle.

PRECIPITATION

EVAPORATION

OCEANS: 97.4%

ICE: 2.6%

◄ Buried treasure
Most of Earth's fresh water lies beneath our feet, in invisible formations known as aquifers. Lakes and rivers make up the tiny fraction called "surface" water.

Coasts: on the edge ►
Magnets to home building, industry, and recreation, coastal zones also are sensitive nurseries for aquatic life, much of it economically valuable.

LAKES AND RIVERS: .01%

SEE ALSO

Pascal's
liquid assets
48-49

Heat
and weather
70-71

We're the
stuff of stars
136-137

Let's drink
to water!
144-145

Seeking
balance: pH
150-151

Ashes to ashes,
rust to rust
158-159

Polar ice ▶

Only about 3 percent of Earth's water is fresh. Approximately 70 percent of fresh water is held in the polar ice caps and glaciers. Essential to life, water makes up 60 to 70 percent by weight of all living organisms and is intregal to the process of photosynthesis.

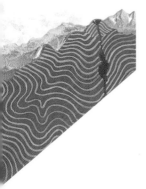

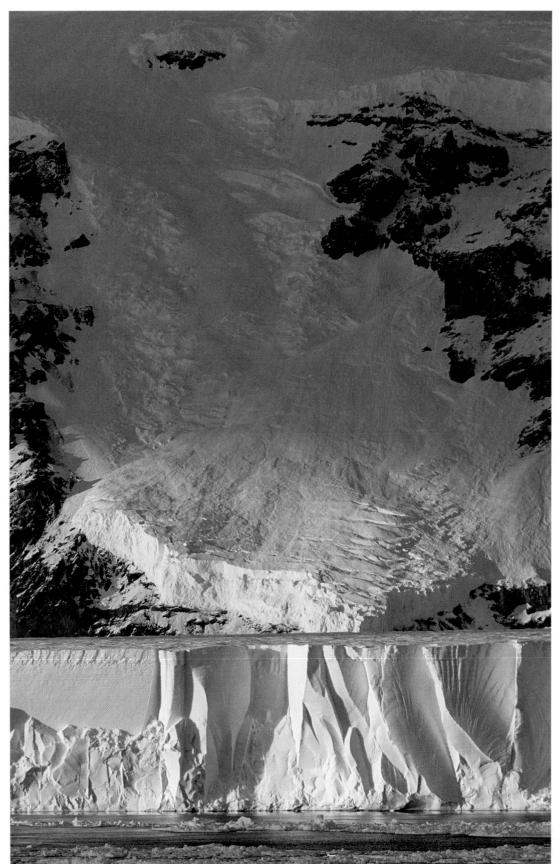

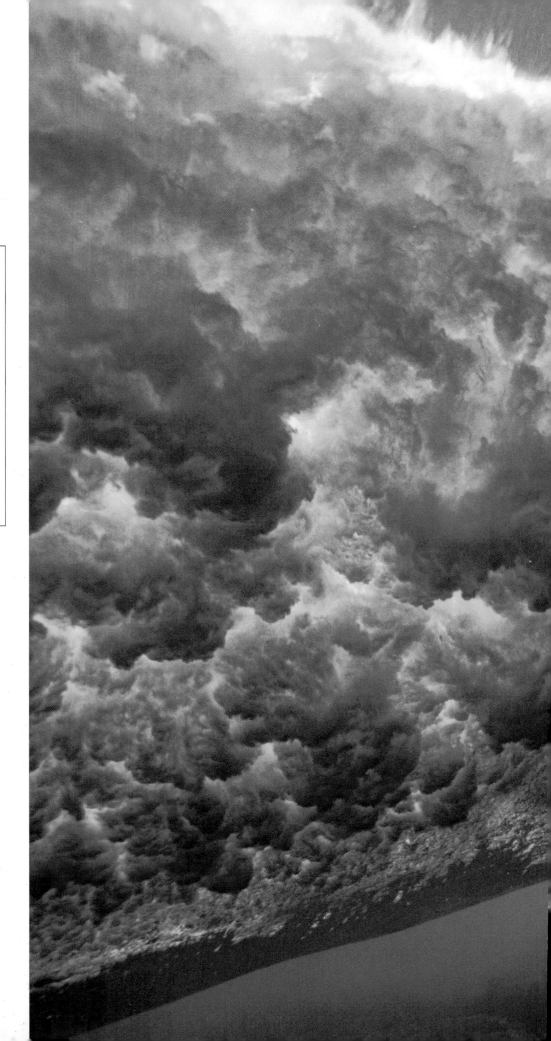

Prime mover

The oceans have a major role in transferring mechanical energy in tides, waves, and currents. As most surfers eventually learn, a cubic yard of water weighs about a ton. Arriving in waves, the sea applies a stupendous amount of force to beaches and shores—and to formerly inland land as the global sea level continues to rise and coastal areas erode. During the 20th century, sea level rose about ten inches, owing to thermal expansion of seawater and gradual melting of ice caps and glaciers. Most projections for this century run from one to three feet.

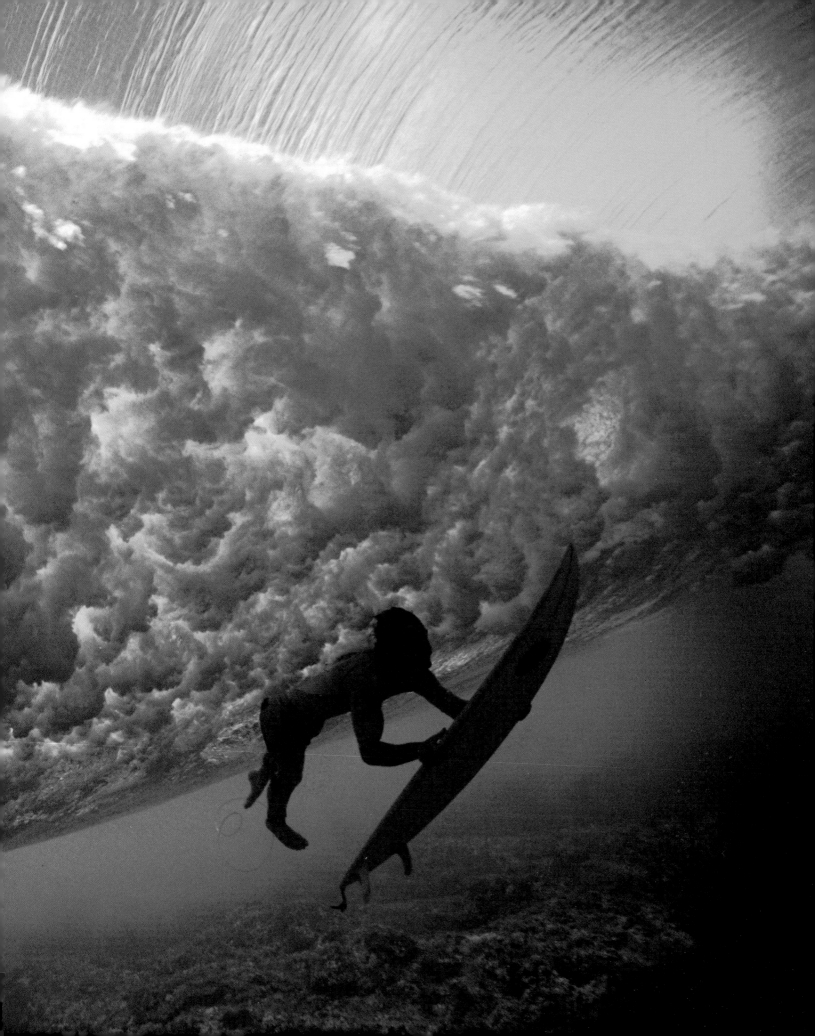

Seeking balance: pH

You may think you left high school chemistry a long time ago, but your stomach—to say nothing of the cosmetics industry—can tell you otherwise.

Deep within you, tiny chemical factories produce hydrochloric acid, or HCl—particularly at mealtime. It and the enzyme pepsin are the primary chemicals for breaking down proteins; the HCl also kills unwanted bacteria that arrive with the meal. Sometimes your stomach's acid production exceeds your digestive needs. That's when you pop an antacid and become a chemist—countering acid with its chemical antidote, a base such as sodium bicarbonate (below).

Acids work by increasing the amount of free hydrogen ions—that is, naked, positively charged protons. The more hydrogen ions, the more acidic the substance. Acidity is measured on the pH scale, which means "power of hydrogen." The scale runs from 0 to 14, with 7 being neutral, the pH of pure distilled water. The lower the number, the more acidic; the higher, the more basic. Bases are the opposite of acids—they're substances that take up free hydrogen ions. Many familiar bases release a negatively charged OH group—called a hydroxyl—which is why their chemical names often end in "hydroxide." In a typical acid-base reaction, the OH from the base combines with the H from the acid to make H_2O and neutralize the solution. That's just what antacid pills do in your stomach. ■

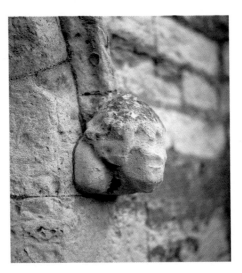

Lakefront view ▶
Swedish workers treat an acid lake with 8,000 tons of lime. The calcium-oxygen compounds counteract the acids that have made their way into the water. But for many lakes, the improvement is only temporary, because acid waters keep flowing in.

◀ Going, going ...
Acid rain contributes to the erosion of stonework throughout the world. This figurehead from a castle in Lincolnshire, England, is barely visible. Elsewhere, carvings have been completely eroded.

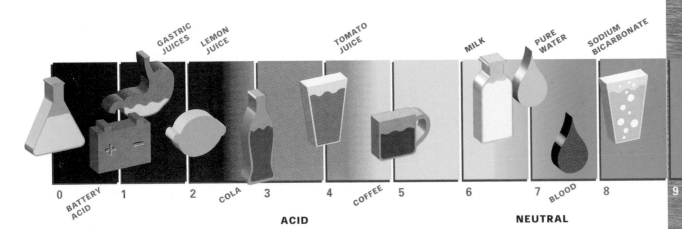

GASTRIC JUICES   LEMON JUICE   TOMATO JUICE   MILK   PURE WATER   SODIUM BICARBONATE   MILK OF MAGNESIA

0   BATTERY ACID   1   2   COLA   3   4   COFFEE   5   6   7   BLOOD   8   9   TOOTHPASTE   10

ACID   **NEUTRAL**   **BASE**

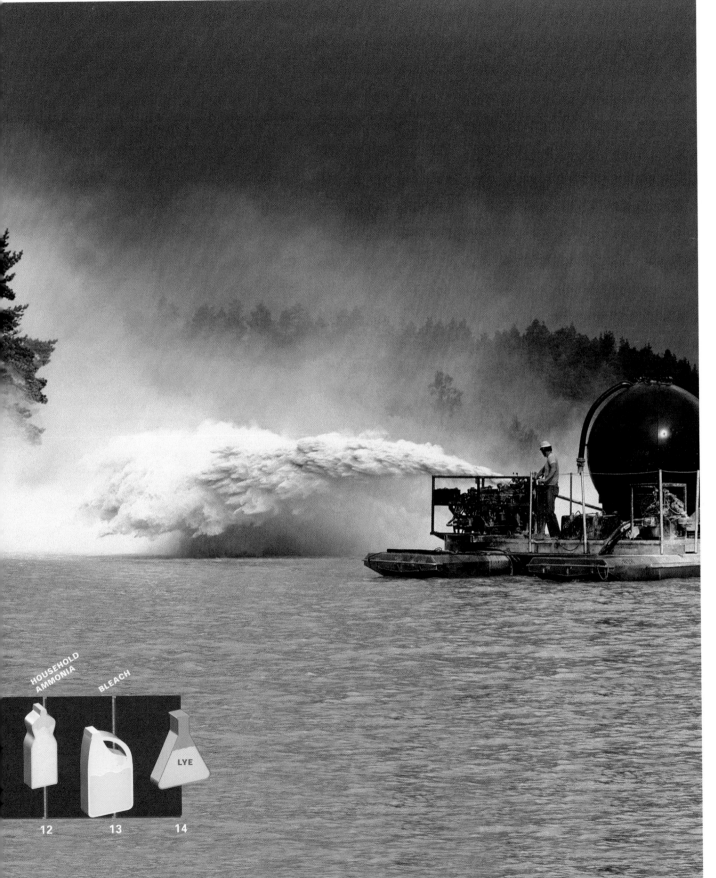

HOUSEHOLD
AMMONIA

BLEACH

LYE

12

13

14

Chemistry for contractors

Civilization never would have risen to current heights if we hadn't learned the basic chemistry of sophisticated building materials.

Human beings seem architecturally determined to reach for the sky, and so far we've gone over a quarter of a mile straight up, with structures such as the 110-story Sears Tower in Chicago and Toronto's 1,814-foot CN Tower. Getting there has required extraordinary ingenuity in finding or inventing materials suited to the challenge. Four kinds of stresses are surpassingly important to construction: tensile (pulling), compressive (squeezing), bending, and shear (twisting at an angle). What you build determines which materials are most apt. If you want to suspend a bridge with cables, you'll want the tensile strength of steel, which is double or more that of nylon. For compressive strength in lightweight structures, virtually nothing surpasses wood. The vertical "grain" texture of its tightly bound cellulose molecules makes it roughly five times as efficient as steel when used as a column.

But beyond four stories high, wood tends to buckle. Stone and brick masonry possess considerable compressive strength but can't bear much tension or shear force, thus these materials have vertical limits. The true high-fliers are two synthetics: reinforced concrete and steel. Used individually or together, they bring the benefits of chemistry to the building industry, offering great strength in resisting all four major stresses. ■

Broken ▶

Interest in skyscraper construction increased drastically after the 2001 attack on the World Trade Center. One focus was on the durability of structures made from steel cables and concrete. Such "pre-stressed" designs, in which the cables are put under tension before the concrete is poured, have proven remarkably strong. This one got through a major earthquake.

High flying ▶

Stone has good compressional strength. But it can't hold up to much torsion or shear—which increases dramatically with a building's height. One solution is to taper the building, à la Egypt's pyramids. Medieval architects used airier devices such as these flying buttresses on the grand cathedral of Notre Dame.

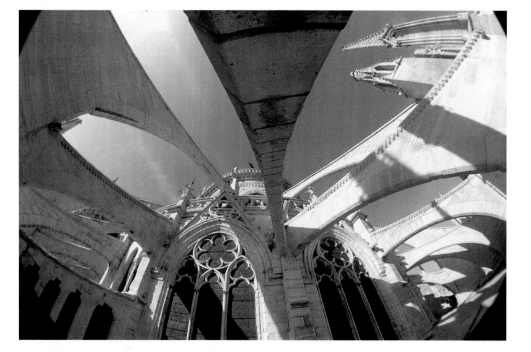

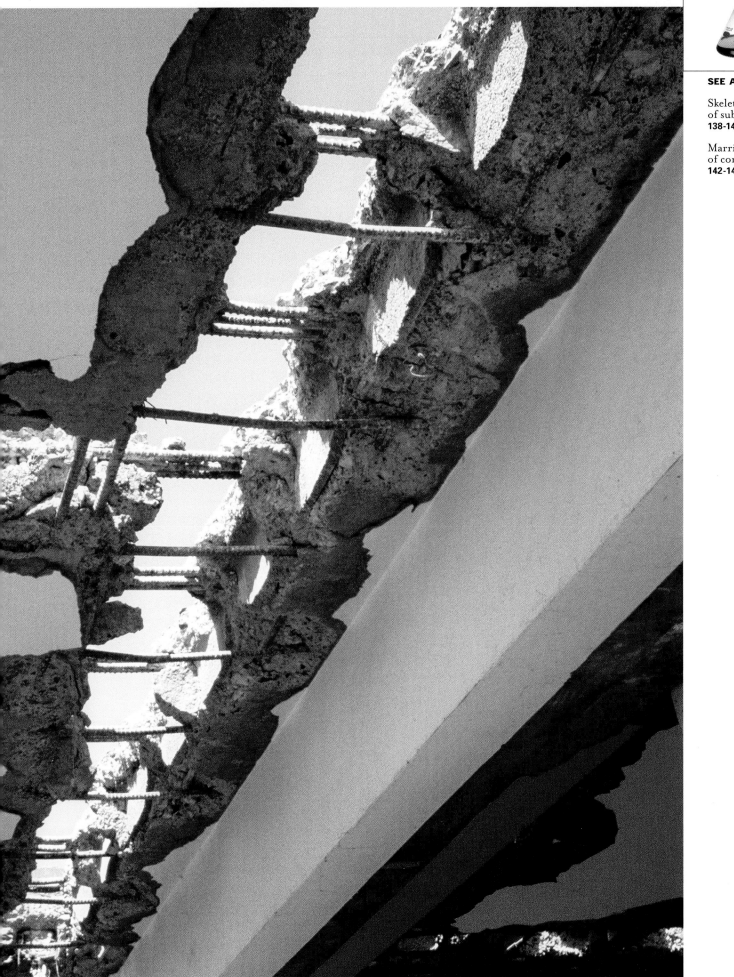

SEE ALSO

Skeletons
of substance
138-141

Marriages
of convenience
142-143

153

Those amazing halogens

We need them. They can kill us. If any single chemical group perfectly embodies the paradoxes of modern chemistry, this is it.

Remember those hyperactive and frequently delinquent children of the chemical family—fluorine, chlorine, bromine, and iodine? They're the ones that lack only one electron to fill their outer shells. They all are intensely reactive, and so are never found free in nature. Due to their frantic propensity to form compounds such as sodium chloride, they are called halogens, from the Greek for "salt makers."

We couldn't live without them. Chlorine ions are essential to dozens of bodily functions. They also help disinfect most urban water supplies. Iodine is crucial to the thyroid gland—and to high-temperature, incandescent halogen lamps. Fluorine, both in toothpaste and in tap water, helps prevent dental decay. Bromine in the form of its silver salt is used today mainly in photographic plates, papers, and films.

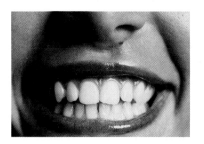

Yet the halogens also are among the most widely feared of elements, for good reason: They are integral to toxic compounds such as dioxins and are the heart of chlorofluorocarbons or CFCs—erstwhile refrigerants and propellants now known to be responsible for depletion of the ozone layer. They also have been implicated in human illness and death. ∎

Molecules with memory

Halogens are film stars, in the form of a silver halide emulsion that coats layers of photographic film. Expose that film to light, and those areas that receive the most light will have the densest concentration of metallic silver. Bathed in developer chemicals in the lab, the silver serves as a catalyst to produce the negative's image. Color film works essentially the same way, but has three layers of silver halide emulsions. Film manufacturers claim that some silver halides are so sensitive that they could detect the light falling onto the Earth from a lone candle on the moon. Of these, the most light-sensitive and most commonly used is silver bromide.

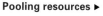

Pooling resources ▶
Chlorine is only slightly less reactive than fluorine. It kills bacteria in our drinking water and swimming pools by ruining the microbes' normal surface chemistry. This same propensity makes chlorine ideal for bleaches—in our washing machines and paper plants—and in certain pesticides and dry-cleaning solvents.

◀ Stronger teeth
A serious overachiever, fluorine is the most reactive element, with tremendous bonding energy. Consequently it can displace just about any other element. That's why sodium fluoride is used in toothpaste. The fluorine atoms combine with the enamel on teeth, making them harder to dissolve.

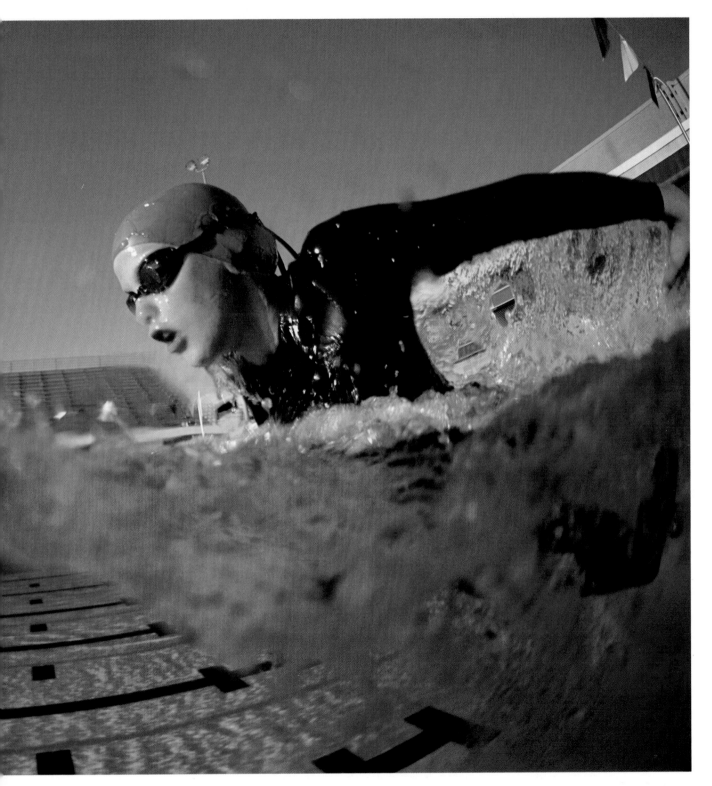

Halogens and the environment

In recent decades we've learned that halogen-containing compounds—even some we long considered harmless—can be risky business.

We're surrounded by halogens everywhere we go. Each of us probably handles dozens of chlorine compounds every day, mostly in the form of plastics made from polyvinyl chloride. Chlorine alone is the tenth most common chemical in industrial production. Taken together, the halogens are incredibly useful. But they also have presented us with unsuspected dangers. More may appear. In any event, bitter debates will continue as we attempt to sort out their adverse effects from their many positive attributes.

Many combinations of chlorine and carbon are profoundly effective as pesticides—and are profoundly hazardous to the environment—because they are both toxic and long-lived. DDT, no longer used in the U.S., is one example; so are the herbicides that were once used to make Agent Orange. Carbon and chlorine also show up in the anesthetic chloroform, in electrical insulators such as polychlorinated biphenyls, or PCBs—now also banned—and in dioxins, a class of more than 200 compounds. Some are deadly, capable of causing birth defects as well as cancers. Most pass through the food chain, build up in fatty tissues, and last for years in the body. They're produced by herbicide manufacture, by paper bleaching, by ordinary power generation, by the incineration of municipal and hospital wastes, and even by forest fires.

Recently, a few kinds of chlorinated compounds have been linked to yet another potential problem: "environmental estrogens." These are substances used in pesticides, or for other industrial purposes, that appear to affect the action of estrogen—a key female hormone—in a wide range of animal species, including humans. Although environmental estrogens are not created by biological processes, there is evidence that they trigger a number of abnormal hormone-like responses in the body. ■

Problem, solution

The earliest refrigerators and air conditioners relied on coolant gases that were either toxic or flammable, so industry devised a seemingly perfect substitute: chlorofluorocarbons, or CFCs. Characterized by central carbon atoms strongly bound to chlorine and fluorine atoms, CFCs weren't toxic, flammable, corrosive, or foul-smelling. Indeed, they were so good that they also served as "blowing agents" to puff up plastic foams into disposable coffee cups, and as propellants for aerosols. But when scientists discovered that CFCs wreak havoc with Earth's protective ozone, the chemical industry responded again—creating a class of coolants that are much less injurious to ozone. The original CFCs, however, are so stable that they can last more than a century in the atmosphere.

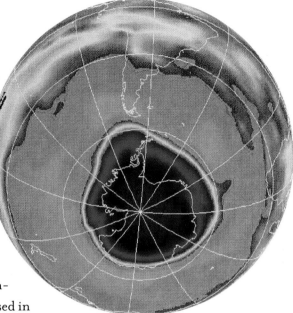

▲ Earth's bald spot
Ozone in the upper atmosphere absorbs a great deal of the ultraviolet radiation that would otherwise cause considerable damage to DNA in numerous species. So when scientists found evidence of drastic declines in ozone concentration around the South Pole, beginning in the mid-1970s, the "ozone hole" became an international crisis. Its appearance was soon convincingly traced to the use of CFCs.

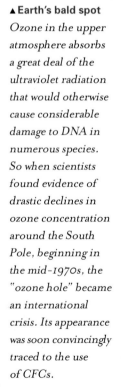

Coolants ▶
Junked cars were once a substantial source of CFCs as their coolants gradually bled away, releasing ozone-destroying gases into the atmosphere. But since the Montreal Protocol of 1987, which severely limited worldwide production of the most worrisome CFC compounds, the ozone-depletion problem at the Poles appears to be improving—very slowly, but surely.

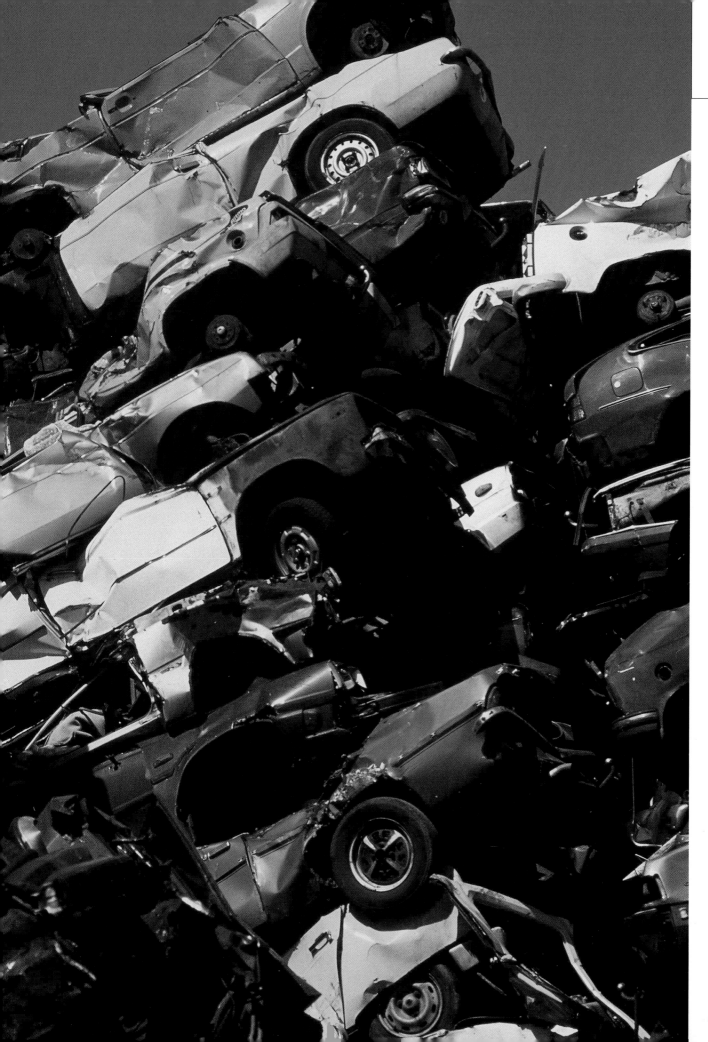

Ashes to ashes, rust to rust

The stuff we breathe in a thousand times a day enables fuels to burn, iron to rust, and a lot of other things to happen. It also keeps our internal fires lit.

hereas halogens are sometimes controversial, oxygen is downright incendiary. This fiercely reactive gas is the most plentiful element in the Earth's outer crust. Bound to hydrogen, it makes water—the most common substance on Earth's surface. Anything termed a carbonate, oxide, silicate, phosphate, or sulfate contains oxygen. So do most things ending in "-ate" or "-ite," as in bauxite (aluminum ore), nitrates (as in fertilizers), and calcite (a form of limestone).

In its gaseous elemental form, oxygen makes up 21 percent of the air we breathe. These days, it is vital to the respiration and metabolism of most animals. But it was a dangerous substance when it made its atmospheric debut some 3.5 billion years ago as a then-toxic waste product of a newfangled biochemical idea: photosynthesis. About a billion years later, the ever more flourishing photosynthesizing plants—especially algae—caused a marked rise in atmospheric oxygen. This led to the oxidizing of rocks throughout the terrestrial surface—a sort of planetwide rust riot—as well as the wiping out of unknown numbers of anaerobic life-forms to which oxygen was a deadly poison. But it also prepared the biome for the advent of new kinds of high-energy creatures who could harness the element's reactive potential. ■

Electrons of rust ▶
Only two electrons short of filling its outer shell, oxygen is always looking for more. The loss of electrons is called oxidation, so iron is oxidized—loses electrons—when it rusts. Oxygen gains them—a process called reduction. Because both occur simultaneously, activities such as rusting are referred to as "redox" reactions.

On the road

Oxygen eagerly grabs electrons from its environs in order to complete its outer shell. The burning of fuels—fossil and otherwise—chemically combines various elements with oxygen, freeing large amounts of energy as heat. That's true in a furnace, a jet turbine, or in a jogger's energy factories, which burn the fuels we call sugars, fats, and proteins. We oxidize sugars and other carbon-based molecules in our cells, releasing carbon dioxide as a by-product and generating the energy our bodies require for activity. The same kind of activity turns the white innards of an apple to an oxidized brown.

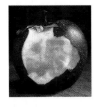

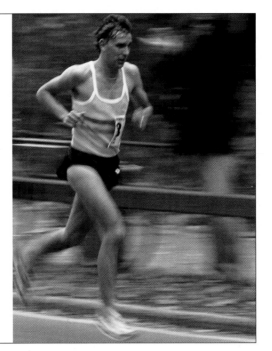

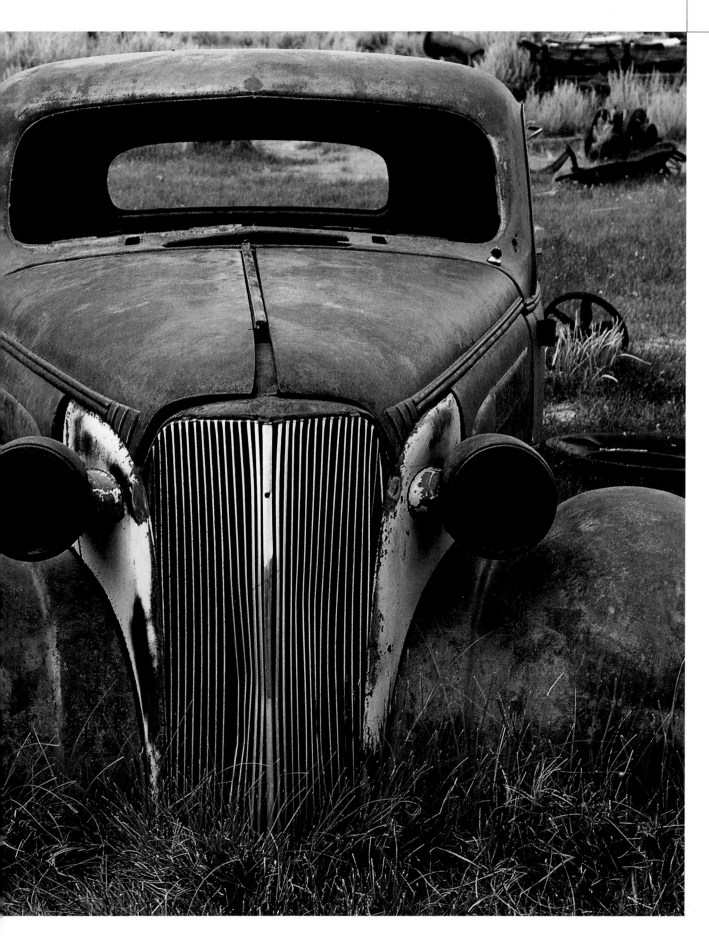

It's what's for dinner

We all need nitrogen, the essential element of any protein. But like thirsty sailors in mid-ocean, we're awash in a sea of nitrogen that we just can't use.

ach breath we take brings in quarts of nitrogen, the gaseous element that makes up nearly four-fifths of our air. Nitrogen also is the must-have ingredient in all proteins and nucleic acids, as well as an important chemical messenger in many body tissues. But despite its atmospheric omnipresence, we supposedly advanced life-forms can't do much with it in its natural, gaseous state; nitrogen's strong, triple covalent bonds make it virtually inert to us, and we simply breathe it back out, unused.

So before nitrogen in the air can be used by living things, it must be "fixed"—that is, converted into compounds with other elements. This utterly indispensable function is performed by the unsung heroes of the biosphere: a few species of "nitrogen-fixing" bacteria. To locate nitrogen that's useful to us, we have to scramble, getting it third- or fourth-hand—after it's been chomped on by the industrious bacteria, and then processed even further by plants and, in the case of meat, by herbivorous animals.

Meat is often, and misleadingly, termed a "complete" food source because it contains all 20 essential amino acids—the building blocks of key proteins—needed for human life. Plants are likewise said to be "incomplete" sources because no single species has the full array. But vegetarians aren't exactly keeling over in the street. That's because a diet based on the right combination of grains, beans, nuts, and other plant proteins will provide all the necessary amino acids—especially if every once in a while you say "cheese." ■

From bacteria to beans to us ►
Nitrogen-fixing bacteria make their metabolic living by binding progressively more pairs of hydrogen atoms to the two-atom nitrogen molecule, until the unwieldy conglomerate divides into two molecules of ammonia—the heart of amino acids. Many of the most familiar of these bacteria live in bulbous root nodules (inset) of plants called legumes. Soybeans and other beans, as well as alfalfa, clover, and acacia trees, all belong to this group.

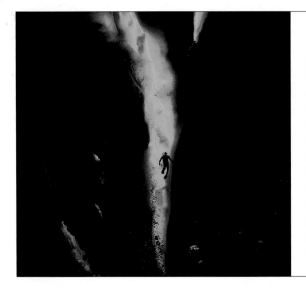

Nitrogen as nemesis

For a descending diver, mounting water pressure causes increasing quantities of nitrogen to dissolve in his bloodstream. The deeper he goes and the longer he stays, the more nitrogen his blood absorbs. High levels can affect judgment much the way that "laughing gas"—nitrous oxide—blurs the senses. This is nitrogen narcosis, also called "rapture of the deep." When a diver ascends too quickly, rapidly decreasing pressures allow the dissolved nitrogen to escape so fast that it may form potentially disastrous bubbles in the bloodstream, resulting in decompression "bends."

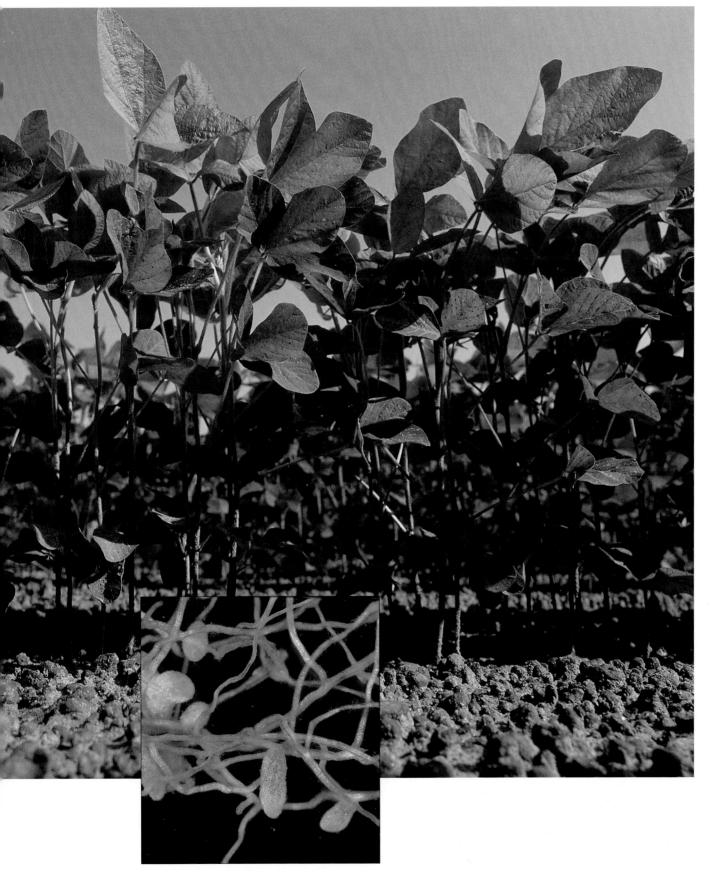

Carbon's black magic

No element is more sociable or versatile than carbon. It is part of some eight million compounds, with more invented daily.

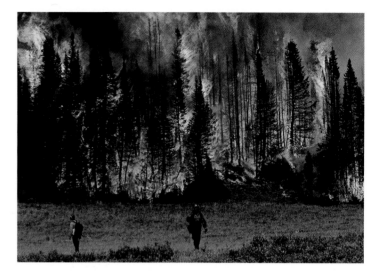

f the elements were to elect a class president, carbon would have to be it. Carbon's astonishing popularity results largely from the fact that its outer electron shell is half full, containing four electrons. That arrangement enables it to form a variety of single, double, or even triple covalent bonds with other atoms. Also, carbon atoms bind readily to each other, creating long chains, branches, closed rings, and a host of combined configurations.

The study of carbon compounds is called organic chemistry because until the early 19th century, scientists believed that only living things—plants or animals possessing some "vital force"—could produce them. That notion was abandoned in 1828 when urea, an organic compound that occurs naturally in urine, was synthesized from lead cyanate, ammonia, and water. Nonetheless, carbon and life are inseparably interconnected on Earth. Whenever exobiologists speculate on the possibilities of life elsewhere, they use as their benchmark the chemical requirements of "carbon-based life-forms," as the biota of our native planet are collectively described. And justly so: Although carbon makes up only .09 percent of the mass of Earth's surface, it's fully 18 percent of your mass. ∎

▲ Smoke to smithereens

After carbon goes up in smoke, what becomes of it? That's a question of burning significance because carbon dioxide is the planet's premier "greenhouse gas," trapping surface heat in the atmosphere and threatening global warming. Of course, we wouldn't be here if CO_2 didn't exist. The planet's surface would be some 60 to 70 degrees colder, on average.

Carbon cycle

Carbon dioxide makes up at most only .04 percent of our air. Yet it is the source of practically all the carbon atoms in every living thing. Animals and plants give back some carbon with each breath. Even soil "breathes" as decomposing organic matter releases carbon dioxide, as part of the complex and endless carbon cycle.

Photosynthesis temporarily traps carbon in biomass; more is dissolved in the oceans or converted into the bodies of marine organisms, which die and sink, taking their carbon out of circulation. Limestone rocks erode, their calcium carbonate surfaces becoming airborne CO_2 and salts. Historically, average temperature and carbon dioxide concentrations tend to rise or fall in tandem.

NATURE'S POWER PLANTS
Sunlight provides energy for photosynthesis, in which plants turn CO_2 into sugars and liberate oxygen.

FUEL IN; CO_2 OUT
A coal-burning power plant pumps soot (solid carbon) and carbon dioxide into the air.

PARK CARBON HERE
Dead, buried vegetation decays, returning carbon to the soil.

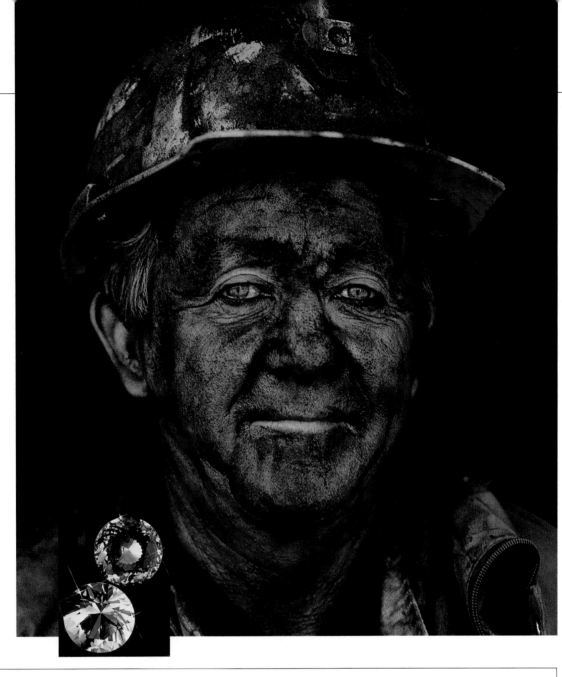

The stuff of diamonds ▶

Carbon is the chief ingredient in coal, which is what deceased vegetation becomes after millennia of heat and pressure have turned it to a sort of rock. It provides 23 percent of the nation's total energy and 55 percent of U.S. electric power. Another form of carbon—diamond—is forged naturally deep underground, by stupendous pressures and temperatures. But now scientists have learned to create various kinds of diamond in the lab— from the sort that are a drill's best friend to jewelry-grade gems of sparkling clarity.

CARBON AND WATER
Rainfall carries CO_2 earthward as mild carbonic acid. This reacts with eroding rock and flushes to the sea, precipitating as limestone.

THE GREAT SINK
Of the estimated five to seven billion tons of carbon we release into the atmosphere yearly, perhaps one to two billion tons are absorbed by the oceans.

ANIMALS INTO STONE
Skeletons of crabs, mollusks, and other shellfish eventually settle to the seafloor and form limestone.

BACTERIA PLAY THEIR ROLE
Microorganisms break down rotting plant material, releasing CO_2 back into the atmosphere.

MYSTERY CARBON
About a billion tons of atmospheric carbon "disappears," going to destinations still unknown.

163

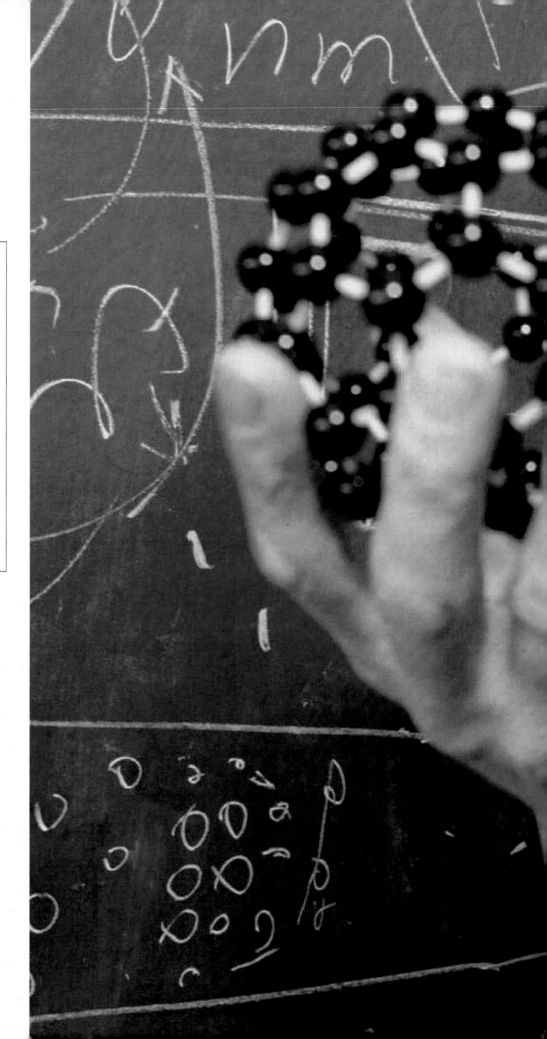

Prized shape

In 1996, three scientists got the Nobel Prize for their 1985 discovery of an unknown and quite unexpected form of carbon: the "buckminsterfullerene" or "buckyball." Under certain extreme circumstances, they found, carbon will roll itself into a soccer-ball-like shape containing 60 or more carbon atoms, bound to one another by single and double bonds. The configuration was given its name by the scientists (including Harold Kroto, shown here) because of its resemblance to the geodesic domes designed by celebrated architect R. Buckminster Fuller. Since then, researchers have created fullerene cylinders—often called nanotubes —that have enormous promise as components of microchips and exotic materials.

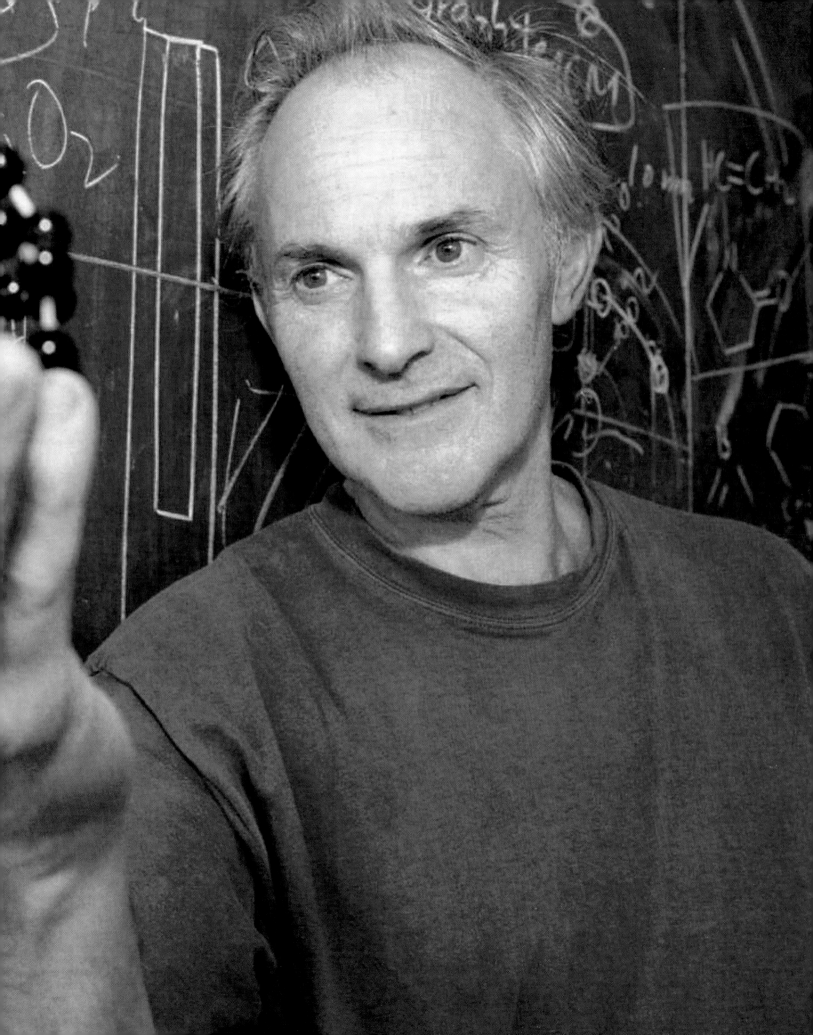

A world of fuels

If you're talking fuel, you're talking hydrocarbons. Nearly everything we burn for energy—from gasoline to the food we eat—is made of them.

Hydrocarbons—chemical compounds of hydrogen and carbon—do a lot more than just power the family sedan. With a couple of additional elements tacked on, they also form the basis for a substantial part of our wardrobes, the bulk of the plastics we use in so many ways, and countless other "miracle" materials of modern life. We deal with hydrocarbons every day. Their seemingly endless variations occupy entire armies of chemists year-round.

Unquestionably the most important hydrocarbons are the fuels, which is why the energy industry is among the world's largest. Another great family of carbon compounds fuels our inner selves. While we eat all kinds of foods, we rely on the carbohydrates—which contain not only carbon and hydrogen but also oxygen—as the chief sources of energy for our cells. Carbohydrates are produced primarily by photosynthesis and consist mainly of sugars and their related polymers: starch and cellulose. In photosynthesis, the leaves of your garden plants absorb energy from the sun to bond water from the soil with carbon dioxide from the air and make glucose, the basic sugar of plant-building and our lives. ∎

Start your engines

Thousands of unwanted compounds make up the glop pumped from the ground as crude oil. Very few are suitable for your car's gas tank. Refineries group components according to their boiling points, which are largely determined by the number of carbon atoms in their molecules' chainlike shapes. The distinctive names of hydrocarbons stem from the configuration and number of carbon atoms. Methane has a single carbon; propane has three; octane, eight. In engines, the shortest and straightest chains tend to burn with an abrupt bang, creating "knock." To rate fuels, engineers chose a chain called isooctane as their benchmark and assigned it a rating of 100. Mixtures that burn more rapidly get a lower "octane" rating; those that burn more smoothly get a higher one.

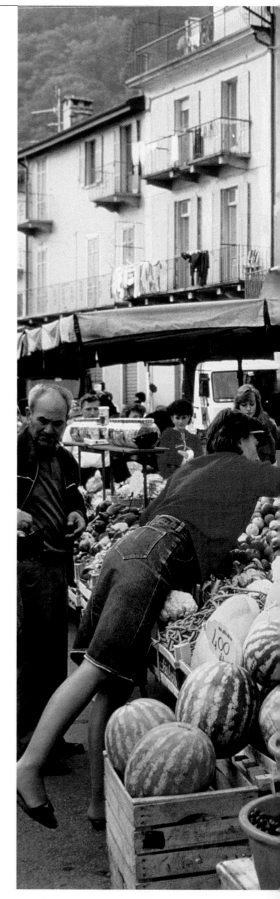

The joys of ethylene ▶
Walk by a produce market, and your nose is assaulted by aromas of ripening fruit. All but hidden in that mélange is ethylene (C_2H_4), a hydrocarbon many fruits produce naturally to stimulate the onset of ripening. Agribusiness uses it the same way: Certain fruits are picked green for long-distance shipping. On arrival, they're dosed with ethylene to trigger ripening.

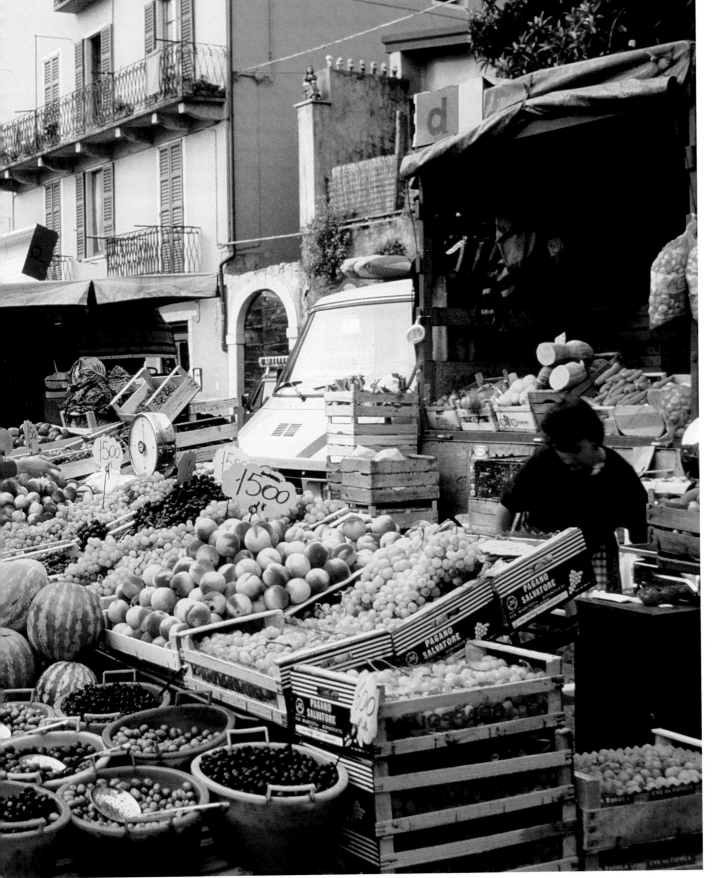

Endless variations on a theme

Once ridiculed as tasteless and cheap, this ever-burgeoning polymer group serves us in so many ways we might properly call this the Plastics Age.

Look around at your clothes, car, workplace, and home. Plastics are everywhere, with more daily as chemists devise new macromolecules that can be millions of atoms long. Whatever their length, polymers usually repeat simple units called monomers, which bind in ways that give each material its unique properties. Polyethylene, used in many plastic bottles, is a soft polymer made of some 50,000 units of ethylene—the agent nature uses to ripen fruit. Plumbing has turned from metal pipes to stiff polymers such as polyvinyl chloride, or PVC: long chains of vinyl chloride, a molecule similar to ethylene but with a chlorine atom in place of one hydrogen. Substitute a carbon-and-nitrogen twosome for the chlorine and you get acrylonitrile—the acrylic fiber found in carpets and some knits. Replace that with a benzene ring, multiply by thousands, and presto: polystyrene. Go instead with two fluorine atoms on each side of a pair of double-bonded carbons, and you've got the monomer for Teflon. Dacron and nylon also are polymers. But for something that truly wears like iron, try one whose units include a couple of benzene rings, some nitrogen, oxygen, and a bit of HCl. It's called Kevlar, and it'll stop a bullet. ■

Web sights ▾

Polymer scientists are at work attempting to produce a synthetic form of spider silk— an astonishing substance that is several times stronger than steel by weight, and even more elastic than nylon.

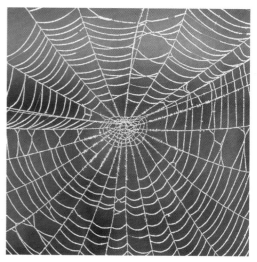

◂**Plastic fashion**

Plastics, in a host of forms, have become an essential part of the apparel industry. For example, polyester fibers can be coaxed into "fleeces" that are warmer than wool but non-absorbent, or into microfibers with the luxuriant feel of suede or linen.

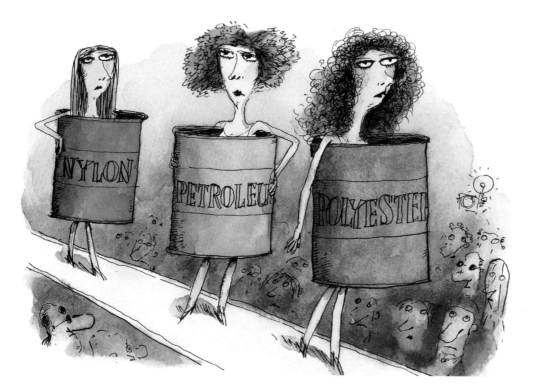

In John's words:
"Many people take plastic products for granted, I guarantee you that I think about plastics every day. The most important part of an amputee's leg is the socket fit. The socket is molded out of plastic and makes a huge difference in the comfort and fit. I don't care if you have a million dollars of equipment connected below the socket, because if the socket isn't fitting properly, you are going to have a bad leg day."

– John Register,
Paralympic long jump silver medalist

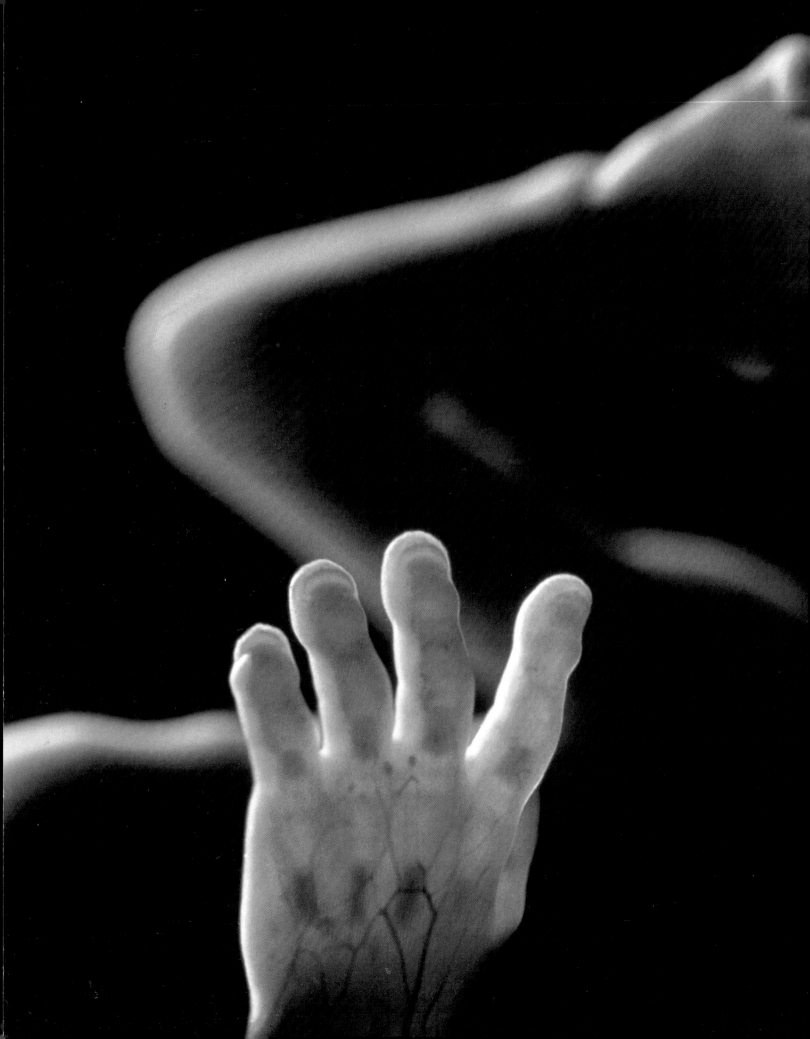

The Chemistry of Life

O ur living bodies are the most complex entities in the known universe. Nothing else—not the white-hot stars that create new elements amid the furious alchemy of thermonuclear fusion, not the mightiest supercomputer ever built—comes close to the wondrous intricacies of the human brain or immune system. And we are but one of millions of species populating this planet, each in its distinctive niche. Modified over billions of years of evolution, life has managed to generate so many diverse permutations—with only a few dozen kinds of basic chemical reactions—that scientists are still hard-pressed to come up with a definition that does justice to all its myriad forms.

What is life, anyway?

Transient yet awesomely varied, life has been around an awfully long time, nearly as long as Earth itself. But we still have trouble defining it.

Three and a half billion years ago, something happened that forever altered the fate of the Earth: A few precocious molecules got together and found a way to make more of themselves. Better yet, they stumbled upon a method of rolling up inside a protective bag that kept their inner workings separate from the encircling brew—thus becoming the primitive forerunners of cells. Life had arrived, and once it set up shop, the planet would never be the same.

Although that intrepid scum was our earliest ancestor, you'd barely notice a resemblance—and not just due to physical appearances. Earth's first creatures were adapted to a very different world. The planet was barely a billion years old, only recently coalesced from a maelstrom of dust and debris around the infant sun. There were no continents, and no oxygen in the air. The atmosphere was a gassy mix of carbon dioxide, methane, ammonia, and nitrogen; it probably looked pink. The surface temperature was perhaps 200°F or more. Comet and asteroid bombardment likely had brought in various organic compounds. Somehow, somewhere, a few blobs of stuff—perhaps resembling modern-day anaerobic bacteria that can live in the hellish environs of deep ocean vents or hot springs—developed the minimal criteria for life: They were able to reproduce and interact with their surroundings. ∎

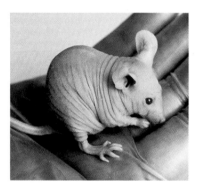

Babes ahoy ▶
Humans have an incredibly long infancy period compared to other animals. Obviously, we have a lot to learn. Yet even tiny babies come equipped with a number of skills, including the amazing propensity to paddle in water.

◄ All living things
have a lot in common. About 98 percent of a human's genetic material is identical to that of a chimpanzee. We share progressively less with more and more distantly related creatures. But plants, fungi, and even bacteria are our kin.

Life's flair for creativity ▶
Over eons, life on Earth managed to evolve into millions of forms, each capitalizing on the advantages of a different ecological niche. This chart arrays major segments of the animal kingdom according to the numbers of species they contain. Insects rule.

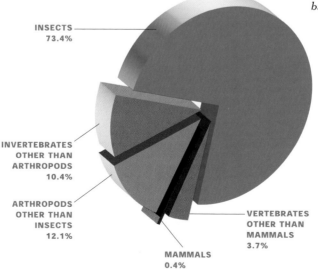

INSECTS
73.4%

INVERTEBRATES
OTHER THAN
ARTHROPODS
10.4%

ARTHROPODS
OTHER THAN
INSECTS
12.1%

VERTEBRATES
OTHER THAN
MAMMALS
3.7%

MAMMALS
0.4%

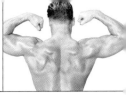

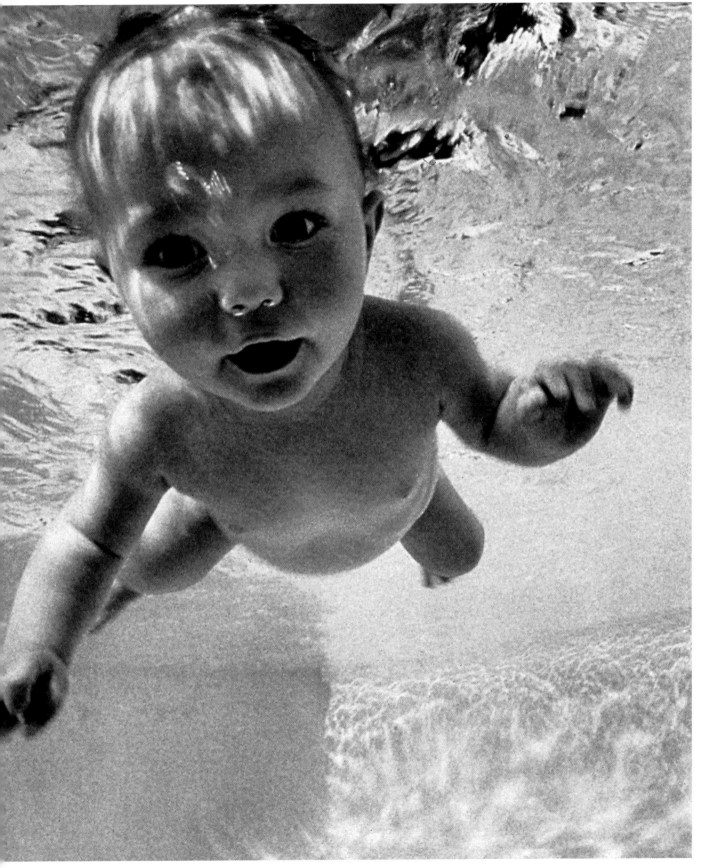

Cells without walls

Unlike plants, all the animal kingdom's diversified members are composed of cells that contain no rigid structural walls—but they do have a lot more.

Any animal cell is a complicated conglomerate, and your body contains some 200 different types. While bacteria are around two microns in diameter—roughly 1/12,000 of an inch—animal and plant cells usually range from ten to a hundred microns wide. Lengths vary far more dramatically, however: Some nerve cells can be as long as three feet.

Life, we believe, first arose in the sea. Even today, many animal cells remain about 70 percent water. The fluid that circulates between them is salty, as is your blood. A normal intravenous "drip" solution contains around one percent salt and five percent sugar. If you're severely dehydrated, however, you may need a hypotonic solution, one with far less salt than normal, to force water into your cells. Because nature is obsessed with equilibrium, water always tries to achieve the same concentration on both sides of a cell's permeable membrane. If there is a greater concentration of water outside a cell, the water will try to squeeze through the membrane in a process called osmosis.

If, on the other hand, your tissues are swollen by water retention—a condition called edema—your doctor might recommend a diuretic. These drugs send chemical signals that increase blood flow to the kidneys or, alternatively, decrease the amount of water and sodium that the kidney tissues release to the blood stream. Either way, urination reduces water volume. ∎

Inside an animal cell ▶

Enclosed in a membrane and reinforced by a network of filaments, a typical human cell contains dozens of specialized components called organelles that are surrounded by gel-like cytoplasm. The central nucleus—a control center stuffed with DNA—has the instructions needed to make thousands of different proteins. When one is required, a corresponding section of the genetic blueprint is temporarily unrolled and a coded version of the protein's formula is recorded on a messenger strand called RNA that carries it outside the nucleus to the ribosomes (1)—which then synthesize new protein from raw materials in the cytoplasm. Proteins that are destined for export outside the cell or that need temporary warehousing or special chemical processing are transferred to the endoplasmic reticulum (2) or to the Golgi apparatus (3) for further work and eventual transport, via containers called vesicles (4), to the membrane. Energy needed for this and other activities is generated in the mitochondria (5), which convert components of food into adenosine triphosphate, or ATP, the basic chemical currency used by all cells.

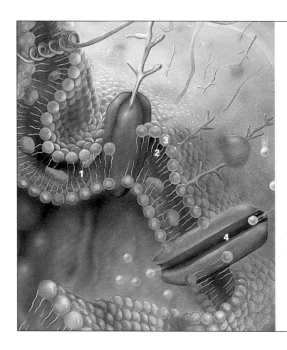

Cell membrane

Each cell deals with the outside world through its membrane (1), a selectively permeable barrier built of two layers of phospholipid molecules. The tail-like end (2) of a phospholipid is attracted to fats but avoids water; its rounded head (3) is just the opposite. In a cell membrane, the fat-loving tails of both layers face each other, leaving the water-loving heads to form the inner and outer surfaces. Tunnel-like channels (4) penetrate the membrane, and many prescription medicines act by modifying membrane permeability. For example, muscle cells need calcium to contract. A heart patient who needs to reduce contraction strength might therefore take a calcium-blocker.

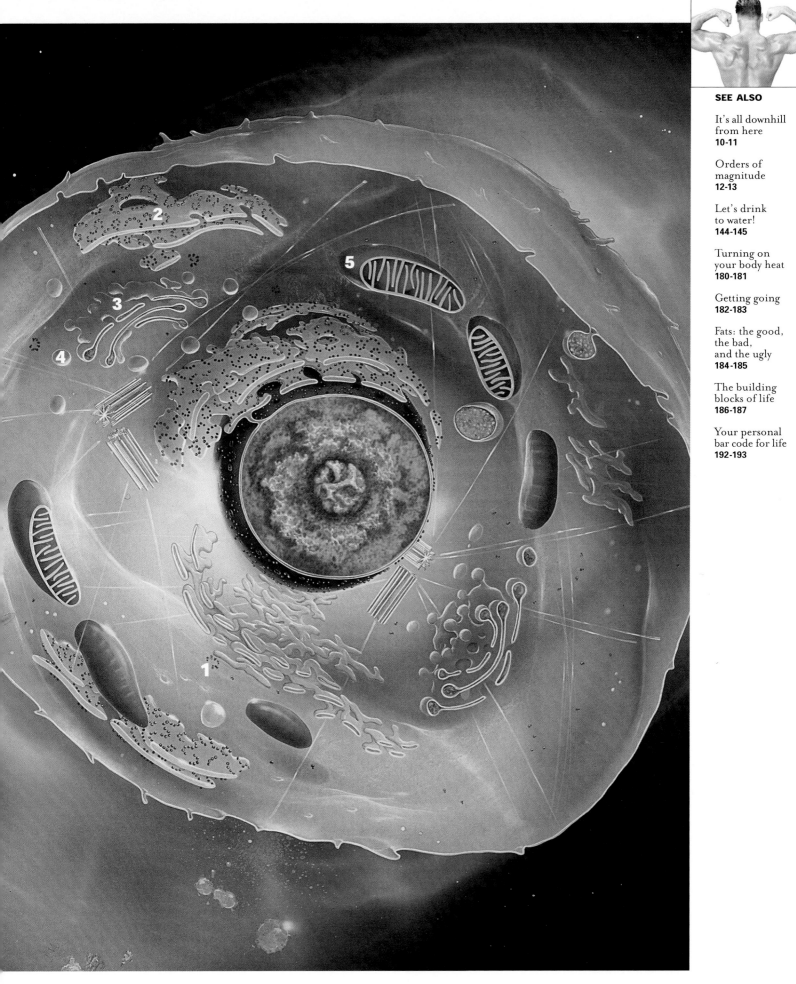

Turning sunlight into food

Plant cells not only feature nifty, self-supporting walls, but they also boast mechanisms to harness the energy of the sun.

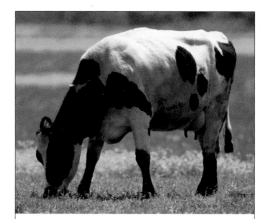

Human beings are what biologists call top predators. That is, we're way up near the top of the food chain. So we don't always remember that almost all living things, ourselves included, ultimately run on sunshine. Animals get their energy by eating plants or other animals that ate plants. Green plants get their energy straight from the sun, using photosynthesis to transform the radiant energy of solar photons into chemical energy, in the form of sugars and other carbohydrates. We're all cruising on sunlight—which is likely to be in plentiful supply for another few billion years or so, before the sun snuffs out for good.

The usual machines of photosynthesis are chloroplasts, cellular inclusions that contain the photon-absorbing pigment known as chlorophyll. Basically, chlorophyll uses sunlight to raise the energy level of electrons so high that they can be detached and passed around through a chain of chemical reactions. The process is complicated, but it boils down to transforming water and carbon dioxide into sugar and oxygen. ■

The circle game

Plants exhale, too, in a process called respiration. It's basically the opposite of photosynthesis. Fortunately for us, plants photosynthesize so much that—even though they do it only while the sun shines—they produce more oxygen and sugar than they consume. Animals benefit from this situation. But we participate in the cycle by breathing out CO_2 while alive and returning to dust when dead.

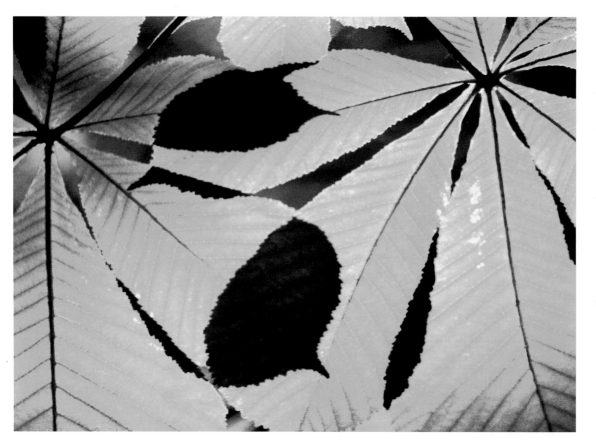

◀ **Delicate balance** *Leaves have to let carbon dioxide in and they have to let some water out. (Otherwise, water from the soil wouldn't flow upward.) But they also have to avoid letting too much water out of their tissues as vapor. Nature's solution is a system of tiny portals called stomata on leaf surfaces that can be opened and closed as needed.*

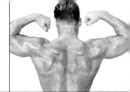

SEE ALSO

It's all downhill
from here
10-11

How heat
gets around
68-69

Ashes to ashes,
rust to rust
158-159

Carbon's
black magic
162-165

Life runs
on sugar
178-179

Turning on
your body heat
180-181

The building
blocks of life
186-187

SUNLIGHT

Food factories ▶

*Within the leaf cells
of green plants,
chloroplasts serve as
the workshops of
photosynthesis—a
series of reactions
that ultimately
produces oxygen and
glucose from water
and carbon dioxide.*

GLUCOSE

O₂

H₂O

CO₂

**Inside each
choroplast ▼**

*Specialized stacks
called grana (below)
contain the necessary
enzymes and other
proteins that use
photoexcited
electrons to convert
solar energy into
chemical energy.*

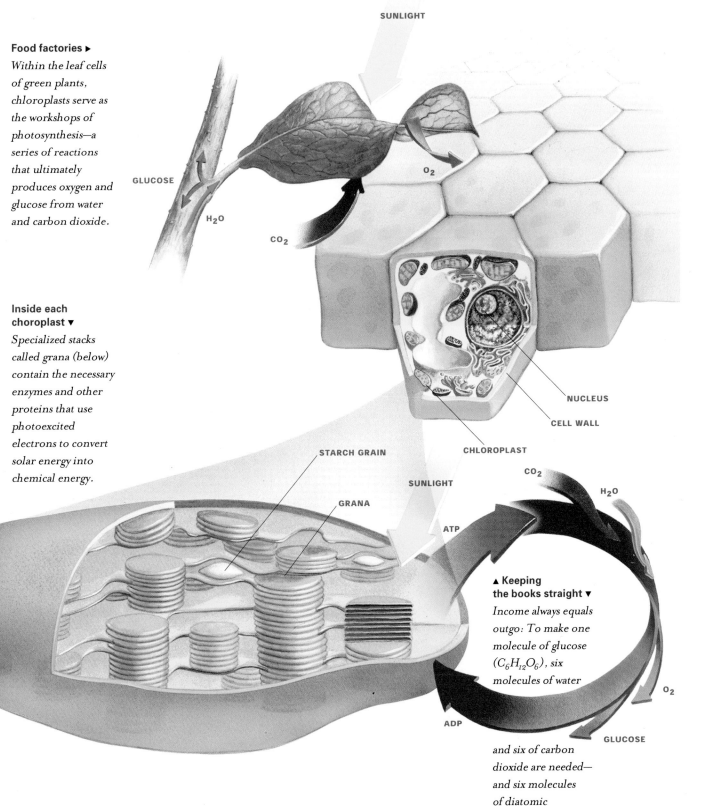

NUCLEUS

CELL WALL

CHLOROPLAST

STARCH GRAIN

GRANA

SUNLIGHT

CO₂

H₂O

ATP

ADP

O₂

GLUCOSE

**▲ Keeping
the books straight ▼**

*Income always equals
outgo: To make one
molecule of glucose
($C_6H_{12}O_6$), six
molecules of water
and six of carbon
dioxide are needed—
and six molecules
of diatomic
oxygen are released.*

Life runs on sugar

The basic fuel for all cells is glucose, one of life's simpler sugars. It's not as sweet as what you put in your coffee, but it pops up nearly everywhere.

Your cells are awfully talented, but most of them wouldn't begin to know what to do with pizza or a cheeseburger. In fact, they won't eat much except the simplest form of sugar, glucose ($C_6H_{12}O_6$). In animals, glucose is also referred to as blood sugar. It's the principal breakdown product of all the various carbohydrates you consume. Glucose is stored by the liver in the form of starchy reserves known as glycogen, which can be converted quickly back into glucose when blood sugar falls too low.

About 60 percent of a healthy diet is made up of carbohydrates, all of which are different combinations of basic sugars and thus are called saccharides, from the Latin word for sugar. Simple ones are monosaccharides; complex chainlike configurations are polysaccharides. You can usually recognize a sugar by its English name: Most end in "ose." So glucose, sucrose, even cellulose are sugar-based molecules.

Plants frequently store sugars in long chains. The most common are starches, which make up the bulk of flour, as well as of most grains. Because starch grains are highly compact units of polysaccharides, they enable plants to pack energy-rich molecules in small spaces. ■

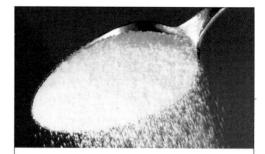

Handedness of sugars

Sucrose, or table sugar, is a chemical combination of two simple, six-carbon sugars: glucose and fructose. Both are highly soluble in water, and both contain the same number of carbon, hydrogen, and oxygen atoms, but in slightly different arrangements. Another name for glucose—dextrose, from the Latin for "right"—refers to the direction it rotates polarized light. Fructose is left-handed. It also is far sweeter, sweeter even than sucrose, and so is being used increasingly by the soft-drink industry.

Sugar buzz ▶
Fruits and some vegetables provide glucose and fructose directly. Others, such as grains, contain sugars in different combinations. Honey is a complex mixture containing about 31 percent glucose and 38 percent fructose (a major component of corn syrup sweeteners), as well as several two-part sugars called disaccharides, a few polysaccharies, and a number of vitamins and minerals.

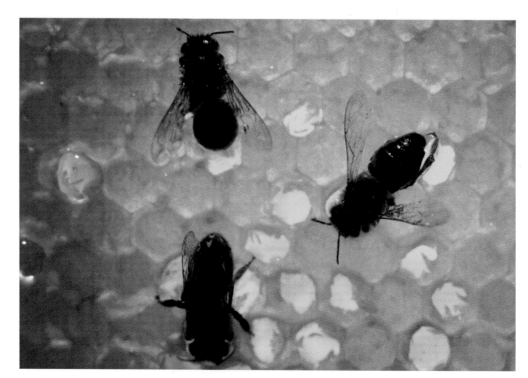

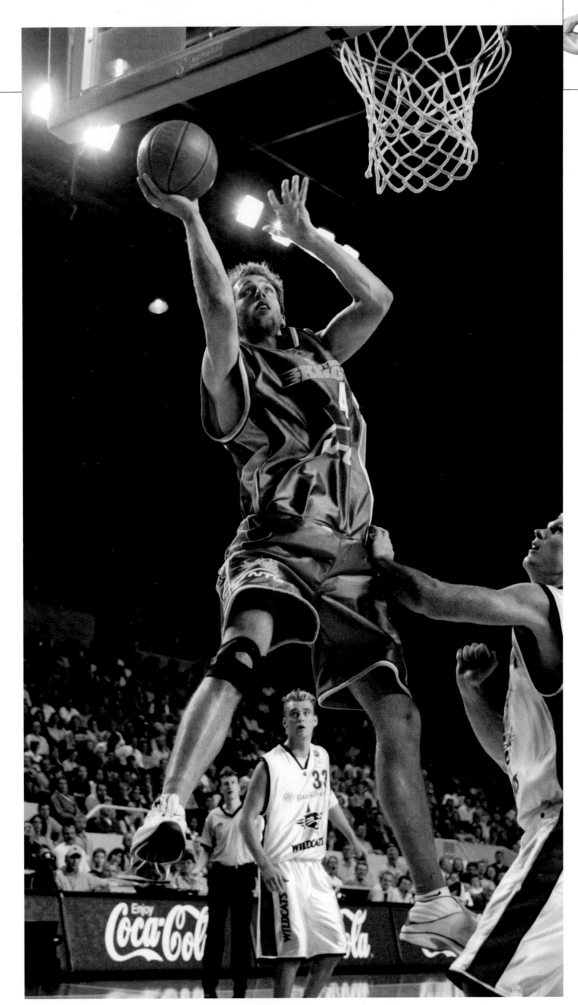

Fueling up ▶

A 150-pound human may consume 450 Calories in an hour-long basketball game—far more energy than is in the amount of blood sugar circulating at any given moment. The liver metabolizes glycogen to replenish blood sugar. One sugar-based molecule we can't digest is cellulose, a complex saccharide found in cell walls and other components of plants. Even ruminants with multiple stomachs can barely handle it. Still, it serves a vital function as dietary fiber, helping scour our digestive tracts.

Turning on your body heat

When cells metabolize sugars and other foods, a fair portion of the chemical energy in those fuels winds up as heat.

ife is a slow burn. What happens when your cells extract energy from sugar is not, in principle, different from what happens when you put a log on the fire. In both cases, a quantity of carbohydrate—glucose in your blood, cellulose in the log—combines with oxygen from your lungs or from the air around the fireplace. Both reactions primarily produce heat energy, carbon dioxide, and water. The standard measure is the Calorie—which has two different meanings. In physics, one calorie (lower case c) is the amount of heat required to raise the temperature of one gram of water one degree Celsius. That's an inconveniently small unit for human energy needs, so when nutritionists use the term, they really mean a thousand times as much: one kilocalorie. That's usually spelled with a capital C or abbreviated as "kcal."

Some of the energy released in breaking down glucose and other nutrients is converted to heat. That's fine, since you need to maintain a core temperature around 100° Fahrenheit for your chemistry to work right. ■

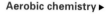

Aerobic chemistry ▶
Occurring only in organelles called mitochondria, aerobic (oxygen) respiration splits glucose molecules and creates molecules of adenosine triphosphate. ATP is the universal energy currency of life, storing energy in its chemical bonds that is liberated when it is "burned" in water.

◄ Anaerobic chemistry
When a cell needs ATP fast, it uses a couple of enzymes and a quick, nine-step chemical shuffle to split glucose in a process akin to fermentation in wine. This anaerobic method exploits very little of the sugar's stored energy, but it does yield two units of ATP in a hurry—without involving mitochondria.

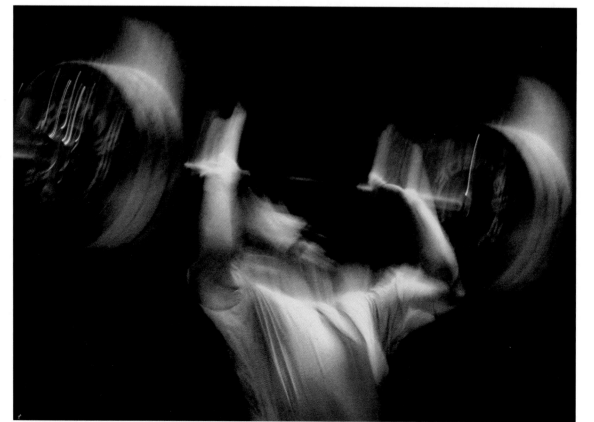

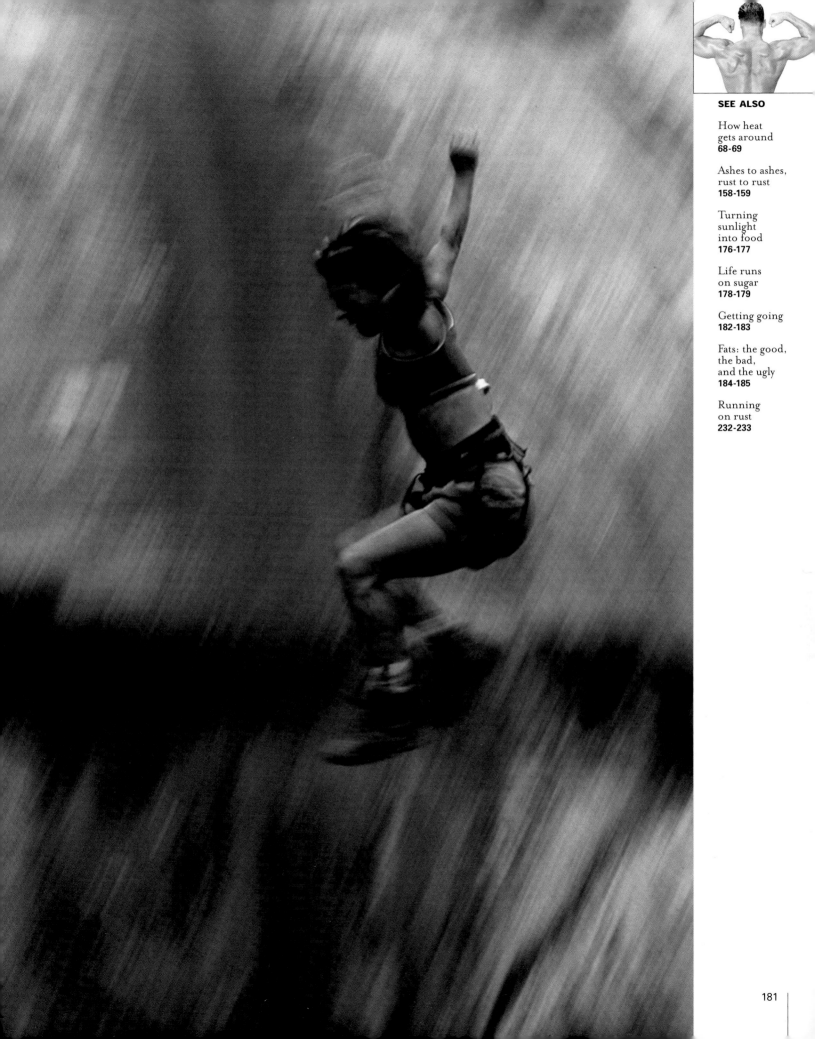

Getting going

How much body fat can you burn while doing an everyday chore? That depends on how intensive the exercise—and on your personal metabolism.

When you move the sofa around the living room, huff up a couple of flights of stairs, or run a hundred-yard dash, you're relying on anaerobic pathways to provide the energy. But when you put down that sofa, catch a few deep breaths, and then take a brisk, substantial walk to your exercise class, your body shifts to its aerobic mode.

The rate at which your body normally consumes energy while at rest—its basal metabolic rate—is generated aerobically. Exercise physiologists often use oxygen uptake—which increases with aerobic activity—as a proxy. The average 150-pound person on an average American diet burns nearly five Calories per liter of oxygen supplied by the lungs. Just lying around in front of the tube, you might suck in some 15 liters of oxygen per hour, which would put your resting metabolism at around 75 Calories per hour. Thus even a dedicated couch potato can consume 1,800 Calories a day without gaining weight. Rates vary, however—from 65 Calories per hour while asleep to about 260 while painting the house or having sexual intercourse, to nearly 600 while jogging and 1,100 climbing stairs.

During exercise, muscles contract when two parallel strands of protein, given a chemical kick by the energy stored in ATP, slide past one another in a sort of ratchet action, causing muscle fibers to contract lengthwise. No two-strand unit ever contracts more than about 30 percent of its total length, which is why weight training and other body-building techniques usually make fibers thicker, not longer. ∎

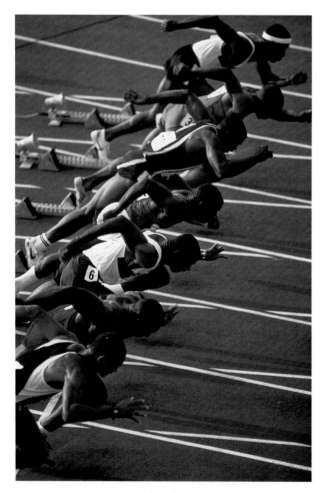

◀ **Some muscle fibers** *are called "fast-twitch." They specialize in anaerobic activity, have few mitochondria but lots of enzymes, and tend to be light-colored. "Slow-twitch" muscle fibers are best in the aerobic mode. They contain many mitochondria, are more resistant to fatigue, and possess a distinctive red color.*

Sweat it out ▶
Up to 75 percent of the energy of glucose is released as heat, some as ATP is created, but most when ATP is consumed by cells. That's why your skin temperature can rise to 104°F during exercise. You shed that heat through conduction, convection, radiation—and in the conversion of liquid sweat to water vapor.

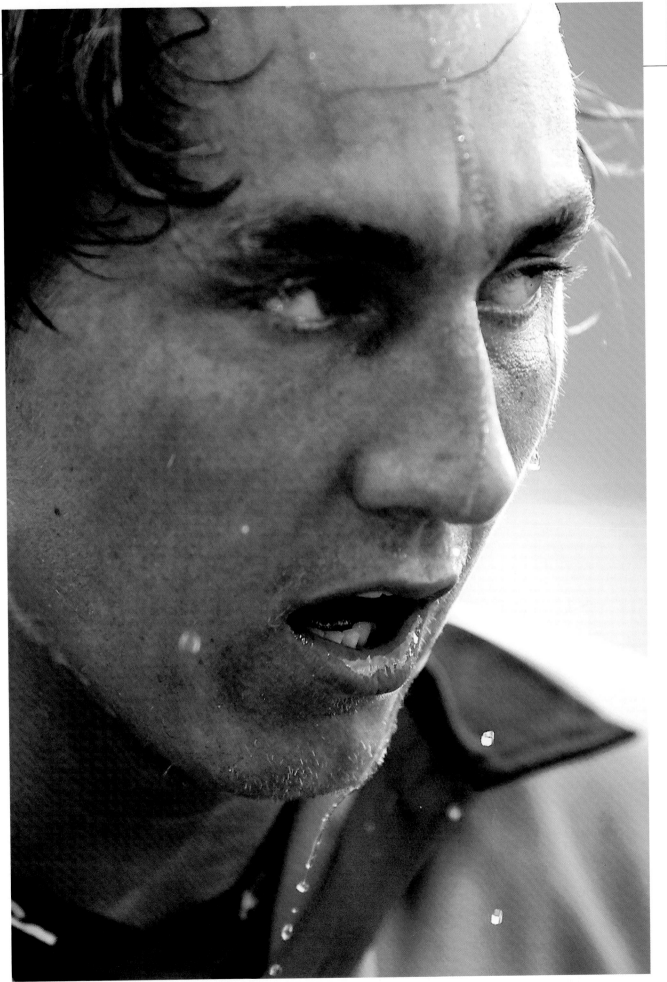

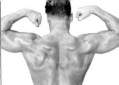

183

Fats: the good, the bad, and the ugly

Fat really makes a lot of sense: It's a compact way for animals to carry their energy along with them. Whether or not they look good is another matter.

n diet-obsessed America, it's hard to say an encouraging word about lipids—a term biochemists use to include all fats and oils. Yet life would be utterly impossible without them. Cell membranes, as we have seen, depend critically on a two-ply lipid layer to keep out unwanted elements. The insulating sheaths of nerve cells are built from fats, which is one reason why growing babies with rapidly developing nervous systems need a fair amount of fat—and why breast milk contains so much of it. Fat makes up only 12 percent of a newborn's weight, but increases to nearly 30 percent by the end of a child's first year. Fats also help generate a number of important hormones collectively called steroids—compounds structurally related to cholesterol—such as estrogen and testosterone. Of course, they also make effective padding to keep you warm and to cover vulnerable body parts such as the kidneys.

Fat city

Everyone has heard dire reports about the cholesterol-boosting hazards of saturated fats, in which each carbon atom on the fatty acid chain is bonded with as many hydrogen atoms as it can hold. In mono- or polyunsaturated fats, the carbon atoms do not have the maximum complement of hydrogen atoms. One unsaturated type, however—"trans" fatty acids, in which the hydrogen atoms are attached on opposite sides across ("trans") the double bond area on the fatty acid chain—appears to have the same worrisome effects as saturated fats.

Although physicians recommend that adults rely on lipids for no more than 20 percent of their daily food intake, most Americans eat a lot more fat—chiefly from meat, eggs, cheese, assorted fried "junk foods," and various plant sources such as nuts, peanut butter, and vegetable oils. Dietitians now recommend that men have no more than 15 percent body fat by weight, and women no more than 20 percent, although social custom has dictated spectacularly different ratios over the centuries. Like sugars, fats are made up of carbon, hydrogen, and oxygen. They are great at storing energy: Pound for pound, fats have more than twice the energy content of sugars or proteins. Most dietary fats are in the form of triglycerides, composed of a ubiquitous alcohol molecule called glycerol attached to three chains of fatty acids, which stream off like tails. Digestion detaches the fatty acids, then metabolizes them and the glycerol; any leftover strands are slated for disposal. If all the carbon atoms in a fatty acid chain are tied to each other by single bonds, that fat is said to be saturated. Unsaturated fats contain one ("mono"-) or more ("poly"-) double bonds between carbons, and are more readily metabolized. ∎

Bulk storage ▶
Your body tucks fat away in pockets of tissue or in special storage cells. Such fat, which contains up to 90 percent of your body's energy stores, is arranged near your center of mass for easy transport. Fortunately, fats and their storage cells are comparatively light. Muscle weighs much more and contains more water: about 75 percent or so, compared to only 10 percent for fats. But every lipid bit helps for these wide-body sumo wrestlers, who can exploit their copious volume in the ring. And, of course, use it to store fat-soluble vitamins A, D, E, and K.

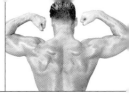

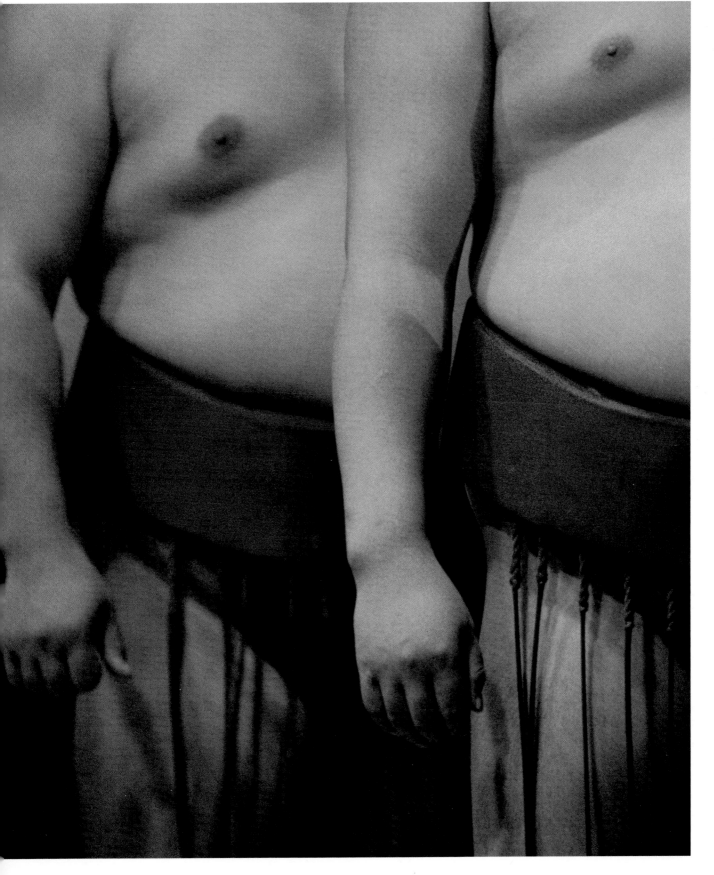

The building blocks of life

From beefsteak to the hair on your head, proteins are pretty special. They give us shape and structure, enable motion, and let body parts talk to each other.

t may not look like it in the mirror, but about half the organic material in your body is protein. Although often associated with muscle, protein is deployed in a stupefying variety of ways and in tens of thousands of forms. All these variations are built from only 20 or so different building blocks called amino acids, arranged in various numbers and sequences.

Like carbohydrates and fats, amino acids contain carbon, hydrogen, and oxygen atoms. Unlike them, they also contain nitrogen—typically in a nitrogen-hydrogen cluster of NH_2—and, sometimes, sulfur.

Of the 20 different amino acids essential to life, your body can fabricate only 12; the remaining 8 must come from your diet, which should be about 15 percent protein. Usually the body does not use proteins as an energy source, although in a pinch the liver converts amino acids into glucose—when instructed to do so by the adrenal glands. That's how you survive prolonged starvation, metabolizing bodily proteins after all available fat and other stored supplies have been used up. Normally, however, proteins and their amino acids are far too versatile to be consumed just for calories. They are the stuff not only of muscle but also of enzymes, antibodies, many hormones, and a host of other essentials.

Every day, your body builds thousands of proteins from 20 basic building blocks. Some become muscle and sinew, some serve as hormones, the chemical messengers that spark growth, tell various organs to speed up or slow down, direct nerve impulses, and—in the case of insulin—help control how cells handle blood sugar. Proteins such as hemoglobin ferry substances through the blood and tissues. Still others make up antibodies, which attack viruses and bacteria. Contrary to what you may have heard, it is not necessary to eat meat in order to obtain all essential amino acids. Vegetable combos can do the job. ■

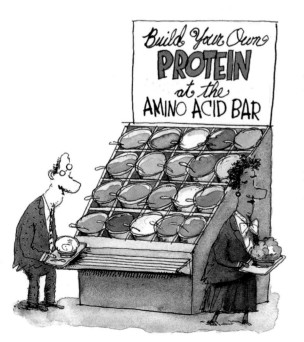

Proteins, proteins everywhere ▶
In fibrous form, proteins create the structure of many body parts. Collagen, for example, adds strength to tendons, cartilage, and joints. Keratin makes your skin tough and waterproof and is the main material in human hair (right) and nails. Different proteins, interwoven with a complex sugar called chitin, form the exoskeletons of numerous hardy critters such as cockroaches.

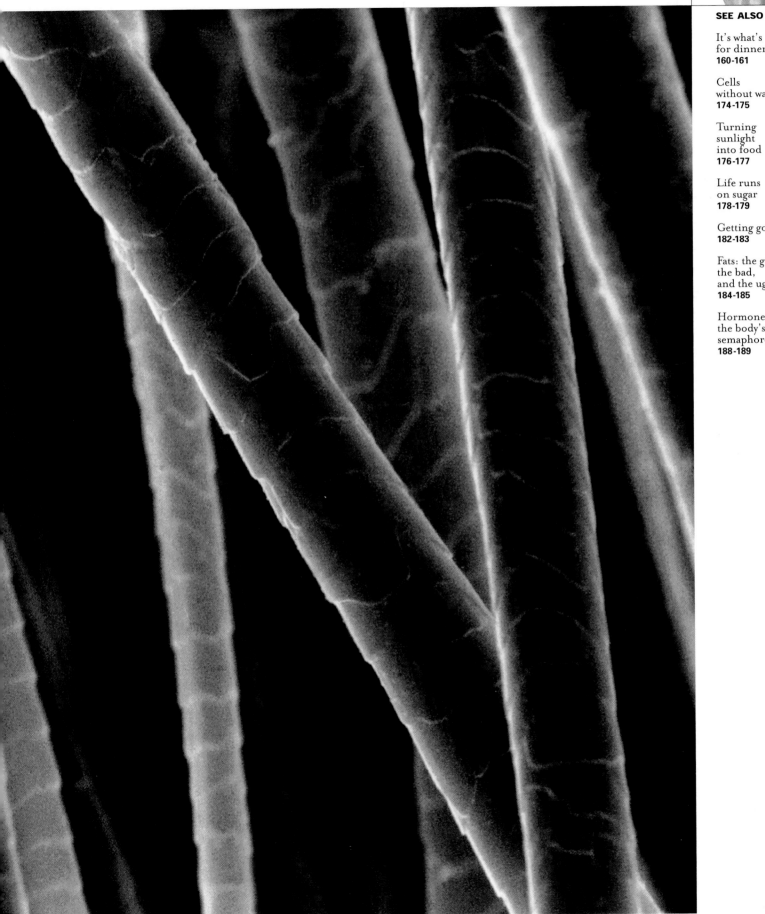

Hormones: the body's semaphores

Your body uses these specialized proteins to turn its different parts on and off. Feedback makes such intrapersonal communication a two-way street.

Mention the word "hormone," and most folks think of sex. That's understandable, since some hormones—notably estrogen and testosterone, the principal female and male sex hormones—play an enormous role in regulating reproductive activity and even in shaping brain structure.

But there are scores of other human hormones, and you'd never make it through the day without them. Each has a specific molecular shape that corresponds to a particular kind of receptor—a sort of docking site on the cell membrane. Each cell has from 2,000 to 10,000 such receptors, and when the right chemical shape locks onto the right receptor, it flashes a signal across the membrane and kicks off some activity within the cell.

Your body's premier collection of hormone-producing glands is called the endocrine system. Its chemical messengers go directly into the bloodstream. Some regulate the production of other hormones as part of a feedback loop. For example, the level of thyroid hormone in the body is constantly monitored by sensors in the hypothalamus and the pituitary gland. When it rises—nudging your whole metabolism into overdrive—the hypothalamus and pituitary produce less thyroid-stimulating hormone, which in turn tells the thyroid to lighten up. Insulin and its companion hormone glucagon regulate glucose levels. When blood sugar rises, insulin from the pancreas speeds up the cellular uptake of glucose. When blood sugar drops, glucagon secretions instruct the liver to break down glycogen into more glucose. ■

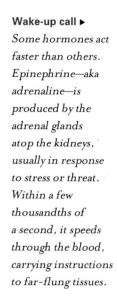

Wake-up call ▶
Some hormones act faster than others. Epinephrine—aka adrenaline—is produced by the adrenal glands atop the kidneys, usually in response to stress or threat. Within a few thousandths of a second, it speeds through the blood, carrying instructions to far-flung tissues.

Replacement hormones

Women are born with a finite number of ovarian follicles, in which eggs develop. When those are used up, menopause and its symptoms—which include decline in estrogen levels—are inevitable. Only a few years ago, it seemed reasonable to attempt to alleviate the symptoms by taking estrogen and progesterone. More recent research, however, has shown that hormone replacement therapy is linked to higher risk of breast cancer and may even increase the incidence of heart attack.

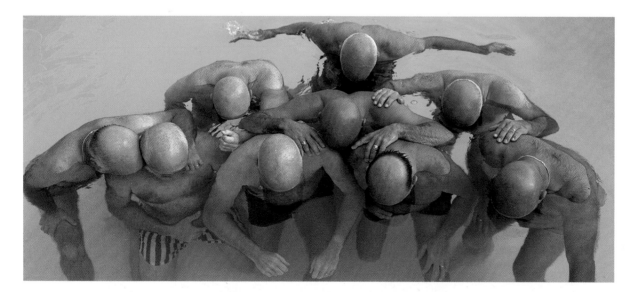

◀ More than sex
Testosterone contributes to male-pattern baldness, as well as to many aspects of metabolism and muscle activity. As a boy's body develops, testosterone causes the larynx to enlarge and thickens the vocal cords, usually resulting in a deeper voice.

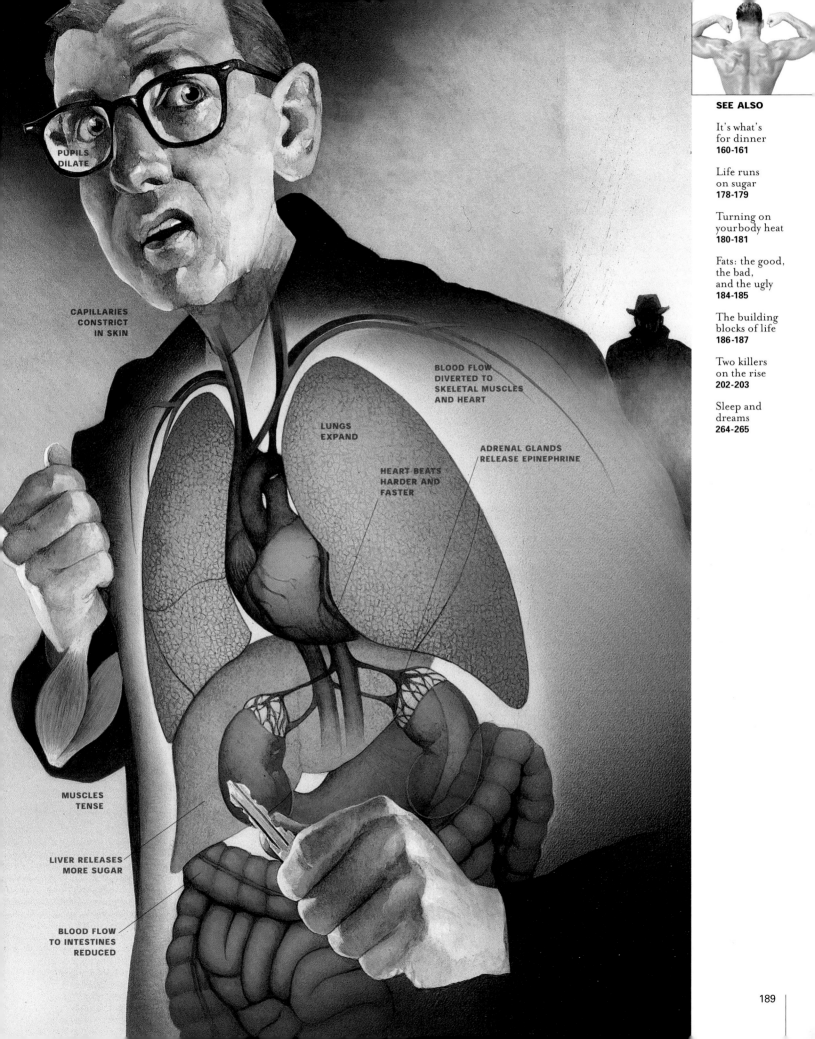

PUPILS
DILATE

CAPILLARIES
CONSTRICT
IN SKIN

BLOOD FLOW
DIVERTED TO
SKELETAL MUSCLES
AND HEART

LUNGS
EXPAND

ADRENAL GLANDS
RELEASE EPINEPHRINE

HEART BEATS
HARDER AND
FASTER

MUSCLES
TENSE

LIVER RELEASES
MORE SUGAR

BLOOD FLOW
TO INTESTINES
REDUCED

How cells change as they stay the same

We all begin life as a single cell. How does that simple entity give rise to brain, blood, skin, liver, eyes, and all our other specialized tissues?

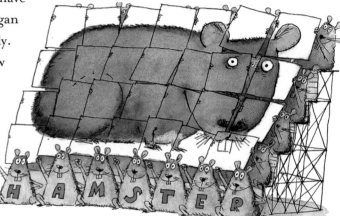

Cells are pretty wonderful things, but they're not especially intelligent all by themselves. So how do they know whether to become muscle or bone or blood instead of something else? Biologists still don't have definitive answers, but they're closing in on the causes. You began existence as a single fertilized egg that divided in half repeatedly. Through the stages of replication from two to eight cells, each new cell was identical to all others. Any one of them could have been physically isolated from others in the cluster—and it would have developed separately, into an identical sibling. Yet after only a few dozen hours of cell division—while your embryo was traveling down your mother's fallopian tube, long before it began to embed itself in the uterine lining—its various cells were already irreversibly committed to becoming totally different kinds of tissues.

Prior to this stage, numerous cells in the evolving embryo had the potential to become any kind of tissue. These "stem" cells may have extraordinary promise in medicine, replacing defective tissues in the brain and elsewhere. ■

Mechanics of mitosis

Most of the cells in your body replicate in a process called mitosis: All genetic information in the parent cell is duplicated and the cell splits, creating two identical versions of itself. In this simplified representation of that process, the cell has only four chromosomes instead of the normal human complement of 46. (Two copies each of the 23 chromosomes containing all 30,000 genes.) Structures called centrioles divide and form filaments as they drift to opposite ends of the elongating cell. The nuclear membrane dissolves. Every chromosome splits into two identical strands; half of each pair is drawn along the filaments toward opposite centrioles. When a complete set of genetic information reaches both ends, a muscle-like contraction squeezes off the center, and two cells form where only one had been.

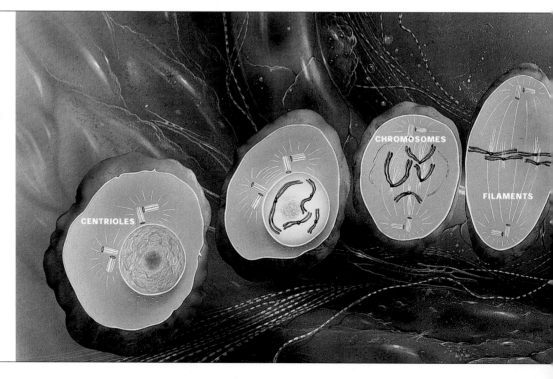

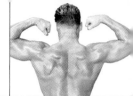

◀ **Cells for sex**
*The difference
between males and
females starts small.
Every embryo
contains 22 pairs of
chromosomes that
are not apparently
sex-related. The
23rd chromosome,
however, consists
of either an XX
combination
(female) or an XY
(male), with each
parent contributing
one of the two. Thus
the father's sperm
determines sex.*

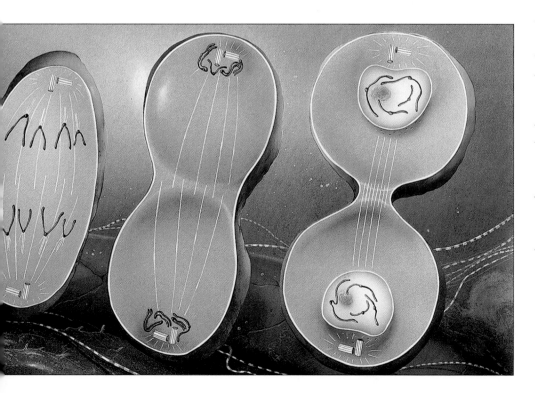

▲ **Cell
differentiation**
*Cells commit to final
form partly in response
to chemical signals
from neighbors,
gradually forming
structures. Those
signals can even be
lethal. The fetal
human hand takes
shape when cells
between the fingers
get orders to commit
suicide in a highly
useful process called
programmed cell
death. Worn-out
cells in your gut do
the same thing.*

Your personal bar code for life

We are all unique—biochemically and physically—thanks to a wondrous molecule that is both infinitely varied and committed to staying the same.

Within the nucleus of each of your cells is a three-billion word set of biochemical instructions for building you from scratch. If you could stretch it out, this string of deoxyribonucleic acid, or DNA, would be about three feet long; in fact, it's tangled up like a ball of yarn. Various parts of the string incorporate the 30,000 human genes, each of which tells your cells how to construct one particular protein out of amino acids. Some of those proteins turn other genes on or off; and many activities require two or more genes working in concert in complex ways that geneticists are only beginning to understand. That's one way proteins affect your body's flow of information.

Interestingly, the human genome—about the same size as that of the rice plant—is nowhere near the largest known. Wheat has an estimated 50,000 genes, many of which you share, along with other animals and plants. Nonetheless, your DNA sequence in uniquely yours. Human genes tend to look pretty much alike from person to person, since we all require the same repertoire of proteins to function. But there is substantial distinctive individual variation in long stretches strewn along the double helix. That is the basis of DNA "fingerprinting," used in criminal trials. ■

Same and different ▶
Nonidentical twins come from two different eggs fertilized by two different sperm. But sometimes a single ovum splits, creating two identical twins (right), each of whom carries exactly the same DNA sequence.

The code ▼
The tightly wound double helix of DNA contains four kinds of "nitrogenous bases." All genetic codes rely upon the sequence of those bases, arrayed in pairs that bridge DNA's two spirals. Thus when strands separate, each attracts base pairs in the same sequence carried by its opposite half.

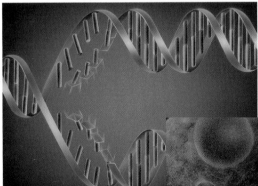

▲ Replicating the code
Because each strand of the double helix is a biochemical "mirror image" of the other, cell division usually results in a perfect transcription of the original DNA sequences. Errors are called mutations.

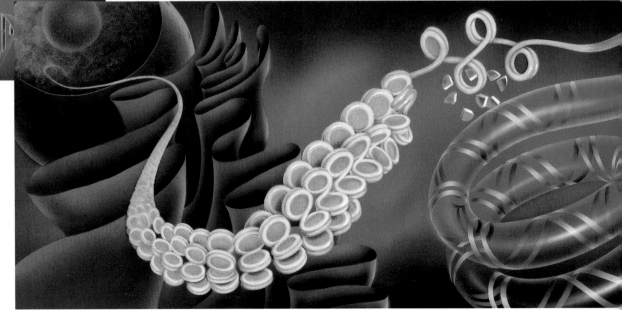

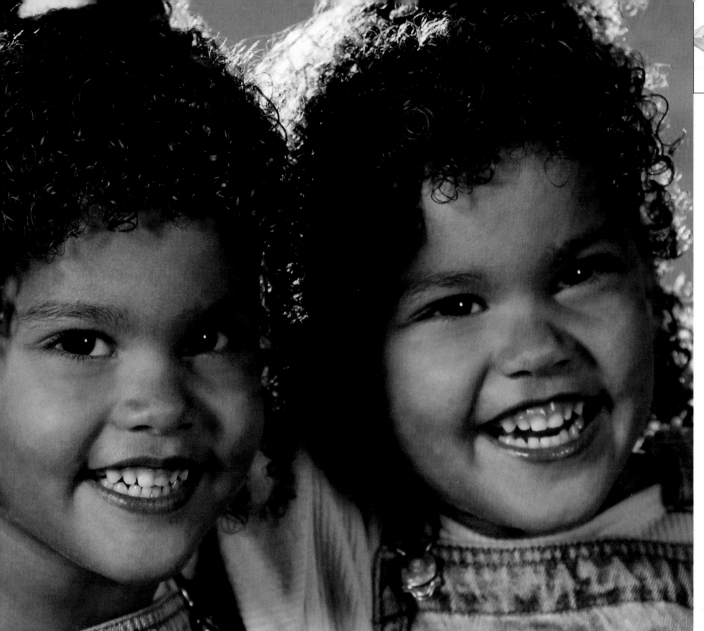

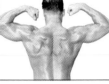

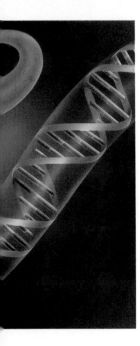

Genetic new deal

The human gene pool retains its diversity because nature has decreed that your sex organs make one kind of cell that doesn't play with a full deck. Ordinary cells contain 46 chromosomes, 23 from each parent. But in the ovaries and testes, egg and sperm are made by meiosis, a different system that ensures incessant reshuffling of genes. Each sex cell, called a gamete, gets only half the genetic material of the cell that created it, and each gets a slightly different selection, since it contains only one gene for every pair in the parent cell. When sperm and egg join in fertilization, the product is a never-before-seen concoction of 23 chromosome pairs. Each embryo then plays this new 46-card hand.

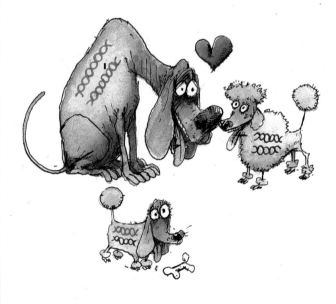

193

Species: It's a family affair

Biologists have a simple rule defining relationships: If two creatures can produce fertile offspring together, they're one species.

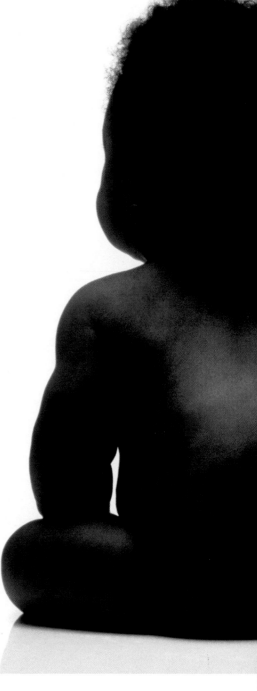

For all that we have genetically in common with mushrooms, muskrats, and marigolds, our nearest evolutionary kin—aside from other members of our own genus and species, *Homo sapiens*—are members of the hominoid family, which includes apes. After that, in decreasing order of relationship, we belong to the order of primates, the class of mammals, the superclass of tetrapods, the subphylum of vertebrates, the phylum of chordates—creatures with a nerve down the back—and finally the animal kingdom as a whole. The key unit of evolution, however, is the species.

That's the level at which genetic changes get made, because only members of the same species can interbreed successfully. Sometimes you can't tell the players without a scorecard. An English sheepdog and a Chihuahua may seem drastically different. But they're both members of exactly the same species: *Canis familiaris.* Ditto for house cats. A fluffy Persian, a svelte Siamese, and a hairless Sphynx are all *Felis domestica,* and any two breeds can make kittens that can reproduce in their turn. Of course, sometimes members of distinct but similar groups can produce hybrid offspring, though not often with true biological success. Thus when horses and donkeys mate, the result is a mule—an admirable and hardy creature, but a sterile one. ∎

Culture collisions ▶
Adaptation to local environments and geographic isolation have led to development of separate, unique cultures, often embodying what appear to be extreme differences among various human groups. Anthropologists are busily studying and documenting those variations as global culture becomes more homogenous.

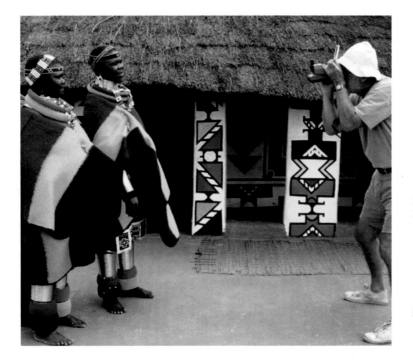

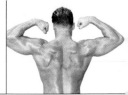

SEE ALSO

What is life,
anyway?
172-173

Hormones:
the body's
semaphores
188-189

Your personal
bar code for life
192-193

Evolution:
nature's test
drives
196-197

We're all the same▼
Over evolutionary
time, local conditions
prompted humans
to develop minor
differences in skin
color and body
structure. But we're
all members of the
same species, which
arose a few hundred
thousand years ago,
probably in Africa.

Evolution: nature's test drives

Though we tend to think of evolution as a steady progression from simple to increasingly complex creatures, it also has some significant pitfalls.

The human body, though attractive enough in its way, is in some respects a pretty shoddy design. Part of the problem is that it's a fairly new one. Unlike many species that have retained and perfected a body configuration over tens or hundreds of millions of years, our hominid ancestors started walking upright only about three or four million years ago. This is hardly enough time to get all the kinks out, and you feel the results every day, chiefly in your lower back.

The human spine derives from a model that evolved to suit quadrupeds—animals that walk on all fours all or most of the time. As a result, we are still getting accustomed to having the full force of gravity directed down our backbones, compressing vertebrae and squashing the disks between. The shock may have been even greater if our ancestors traveled by swinging from the branches of trees before going bipedal, as some paleontologists believe.

Indeed, there are many ways in which modern life is at odds with evolution. For example, we've learned enough about prenatal development to have healthier babies with higher birth weights. But the human pelvis, which governs the size of the birth canal, has not changed to keep pace. As a consequence, say some experts, we have more obstetrical problems and operations than we should—although physician practices also contribute to the increase. ∎

Another failure to adjust ▶

Life was pretty tough back when our ancestors got going. And so was the food: As a result, our jaws have a lot less work to do than they did a million years ago and have shrunk in size and strength. But we still have the same number of teeth—a situation that keeps orthodontists busy.

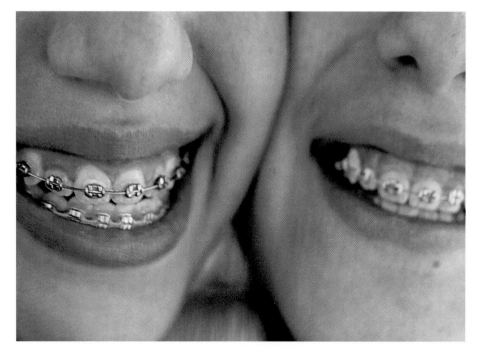

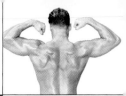

SEE ALSO

Animals at work
38-39

Your personal
bar code for life
192-193

It's a
family affair
194-195

Intimations
of mortality
266-267

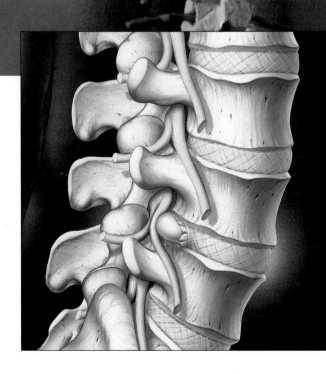

◄ Oh, my aching back

Time seems to have brought some improvements to the human spine: There is evidence that our vertebrae may have gotten thicker at their bases, and we've developed a peculiar but handy S-curve in our backs that helps us stand up straight and sit properly. Still, back pain ranks among the most frequent of complaints, and evolution may never improve the situation— since its onset often occurs after one's genes have been passed on. The pressures of modern life cause the rubbery disks between vertebrae to squeeze outward, or herniate, crimping nerves and causing misery.

The death of a cell

Like idealistic socialists with a suicidal side, cells exist solely for the good of the organism. Yours are dying all the time—just to keep you healthy.

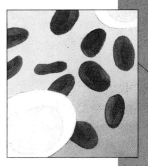

**RED BLOOD CELLS
120 DAYS**

Staying alive means dying a little bit every day. Many cells in our bodies have such miserable jobs, or are forced to work in such hazardous environments, that they have extremely short life spans. Today alone, for example, your liver and spleen will get rid of about a hundred billion red blood cells that are over the biochemical hill, as well as another few billion aging white cells.

By the end of the week, you will have replaced nearly all your stomach lining and much of the inner surface of your small intestine. In contrast, some systems—including nearly all nerve cells and skeletal muscle cells—do not divide once they attain their mature shape. They are expected to last you a lifetime. In between these two extremes are cells with life spans of a few weeks, months, or years. How do they know when to live and when to die? Where is the programming for each cell type?

Although the answers aren't understood in complete detail, it appears that a substantial fraction of the cells in your body are programmed to commit suicide. They carry the equipment to do so in a remarkably tidy fashion, and will use it—unless they are signaled not to by a stimulating factor, a growth hormone, or some other encouraging compound. Absent such signals, it is not obvious why cells should get "old" at all, and the process has baffled scientists since the 1960s, when it was first discovered that most cells can divide only a limited number of times in culture. One mechanism that may affect longevity is the gradual degradation of structures called telomeres at the ends of chromosomes. They're prone to errors and shortening in the course of division, and research shows that cell aging is directly correlated with telomere length. ■

Replacement cells ▶
Like most neurons and the cells of the lens in your eye, heart muscle cells (below) never divide and cannot be replaced. What you start with is all you get for your entire life.

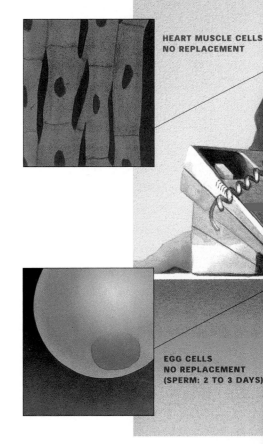

**HEART MUSCLE CELLS
NO REPLACEMENT**

**EGG CELLS
NO REPLACEMENT
(SPERM: 2 TO 3 DAYS)**

Cell suicide

Programmed cell death, or apoptosis, serves many valuable purposes. Sometimes it is simply necessary: For example, the proper formation of circuits among brain cells requires that "surplus" cells die off and get out of the way. Alternatively, if a cell has been damaged by ultraviolet light, invaded by a virus, or injured by oxidative chemicals, then it either kills itself from within, by triggering a cascade of reactions, or places chemical flags on its membrane , marking it for death by the immune system's roving "killer" cells. Scientists are intensely interested in the process of apoptosis. If it can be understood and controlled, the mechanism might be used selectively to urge cancerous cells to destroy themselves. And, of course, it might reveal the biochemical methods for extending the life span of cells in culture—or of entire organisms, up to and including human beings.

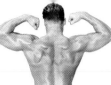

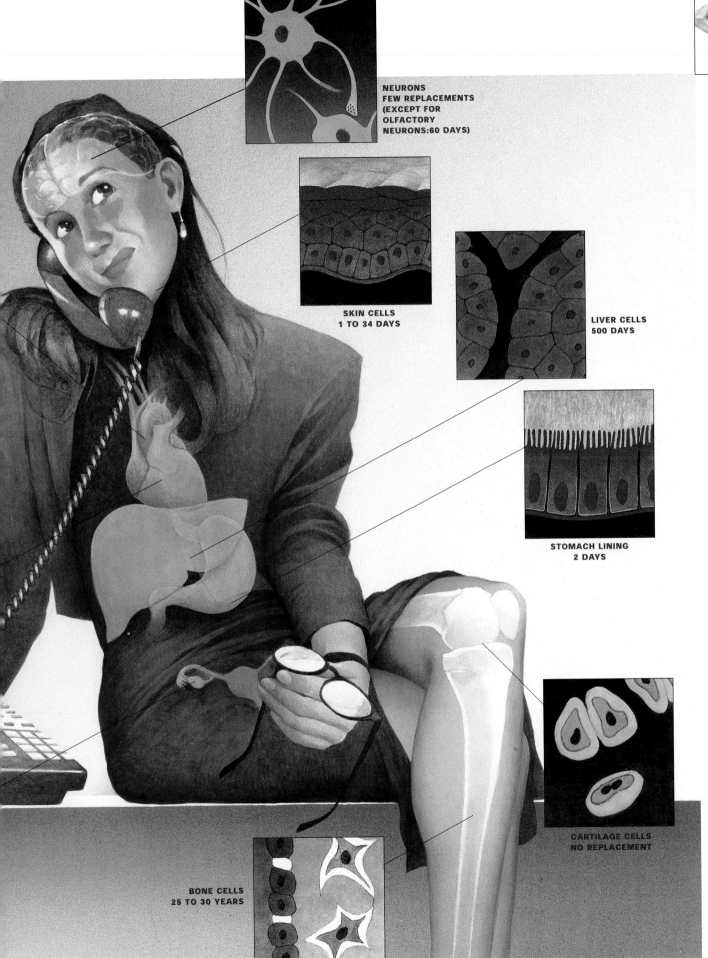

NEURONS
FEW REPLACEMENTS
(EXCEPT FOR
OLFACTORY
NEURONS:60 DAYS)

SKIN CELLS
1 TO 34 DAYS

LIVER CELLS
500 DAYS

STOMACH LINING
2 DAYS

CARTILAGE CELLS
NO REPLACEMENT

BONE CELLS
25 TO 30 YEARS

Cells that never die

Like the Dracula legend, cancers flout the rules of life. Instead of dying a normal death, their cells continue to grow and proliferate indefinitely.

Sometimes the normal processes of cell growth, death, and regulation get completely deranged. Cells are transformed and begin robbing their neighbors' blood supply, like vampires. They defy normal limitations on growth, multiply unceasingly, and then invade space reserved for others. They refuse to die on schedule. These are cancer cells, and they are responsible for about two million new cases of various illnesses and over half a million deaths every year in the United States. One out of five people in industrialized countries will die of cancer, most of them from tumors that originated in a lung, breast, prostate gland, colon, or rectum. And except for a few very clear instances, such as lung cancer caused by smoking or exposure to certain chemicals and radiation, no one knows exactly why.

Unlike smallpox or malaria, cancer is not a distinctive disease with a single cause. It arises from various kinds of mutations in a cell's DNA. Some

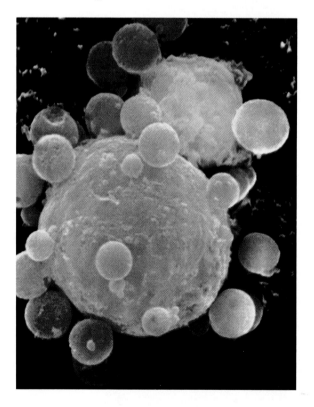

mutations occur naturally. In the course of an average human lifetime, billions of trillions of cell divisions take place. Given this huge number of replications, mistakes are bound to happen. They may result in a disadvantageous change for the organism; some may even spawn cancers. More often, mutations and cancers are prompted by external causes: Carcinogenic substances such as tobacco smoke and various industrial chemicals, ultraviolet radiation and x-rays, perhaps even certain viruses. Doctors assume that as many as nine out of every ten cancers may be due to environmental factors. That might explain why men who immigrate to the United States from Japan or China, where active prostate cancer rates are among the world's lowest, soon experience a rate similar to that of native-born Americans, the world's highest. ■

Cancers ▶

It probably takes several different mutations to turn a cell cancerous, and such miscues usually need to alter one or more DNA sequences that affect cell proliferation. So cancer tends to be a disease of old age.

◀ Homegrown warriors

Some cancerous cells become so different from normal ones that the immune system recognizes them as foreign. Killer T-cells (shown in orange)—a type of white blood cell— then approach such "aliens" (pink) and dispatch them as if they were more conventional threats, such as disease bacteria or cells infected by viruses.

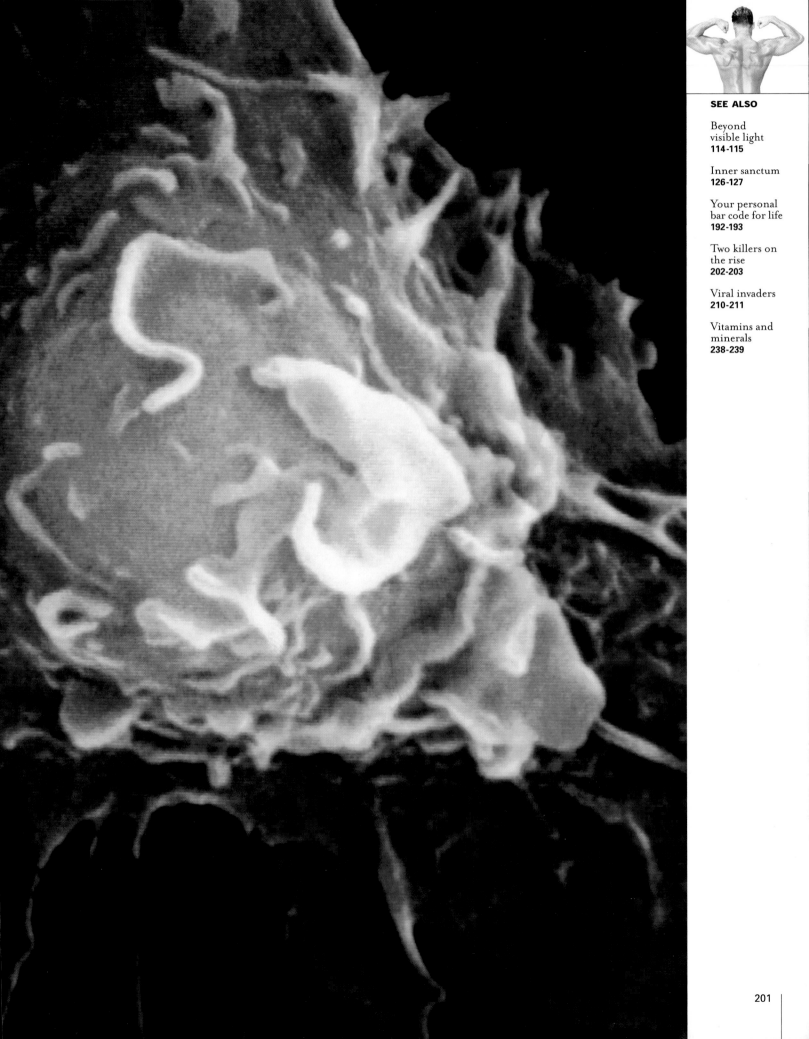

Two killers on the rise

In America and Europe, cancers of the breast and prostate are epidemic—and we don't know why. Yet treatments for both are increasingly successful.

For women, few prospects are more terrifying than breast cancer. For men, the corresponding horror is prostate cancer. Together they account for about one-third of all new cancers diagnosed in the United States every year. The rates for both are increasing, partly due to better detection, but also to reasons that are still unknown. About one in eight American women will get some form of breast cancer, usually after age 40. It is the second most common cancer in women (after lung cancer), and currently kills about 28 in every 100,000 American women. Death rates, however, have declined, thanks chiefly to early diagnosis. Each year about 200,000 new cases of breast cancer are detected, and some 40,000 people die of the disease—including 200 to 300 men.

In the past, researchers have linked increased risk of breast cancer to abnormal alcohol consumption, use of oral contraceptives, and obesity. But the most critical factor seems to be the female hormones, estrogen and progesterone. Women who have children early in life generally have a lower incidence; women who never have children or who have them late tend to have higher rates. This may explain why breast cancer rates are far lower in Asia than in the United States and Western Europe. ∎

Angiogenesis

All cells, including those in cancerous growths, need blood to feed them. So a tumor cannot enlarge or spread unless it is nourished by a network of blood vessels. Scientists are studying the biochemical signals that promote the formation of new blood vessels —a process called angiogenesis—in hopes that they can find a way to reverse it. Many labs are focusing on the sequence of actions that prompt endothelial cells (those that make up blood-vessel walls) to suddenly begin to divide and proliferate. Dozens of therapeutic compounds that stop this action, called anti-angiogenesis drugs, are being tested in human trials.

Tumors of the breast ▶
The most common form of breast cancer arises when tumor cells form in the ducts that connect the breast's numerous lobules. Chances of successful treatment are directly related to the size of the tumor and whether it has metastasized (spread from its original site), so physicians advise monthly self-exams and regular checkups. After age 40, every woman should have a mammogram at least once every two years; after 50, it should be once a year.

Prostate cancer ▶
Of all new cancers found in American men, one-third occur in the prostate, a walnut-size gland that lies just under the bladder. About 220,000 new cases are detected yearly in this country, where the disease's mortality rate has risen considerably since the 1930s. Prostate cancer now ranks second only to lung cancer as the leading cause of cancer death in males.

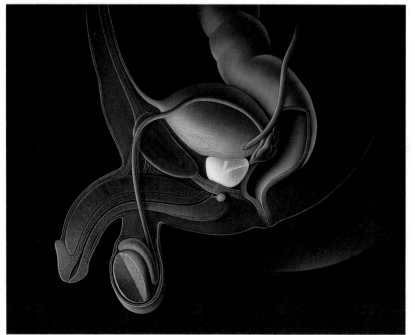

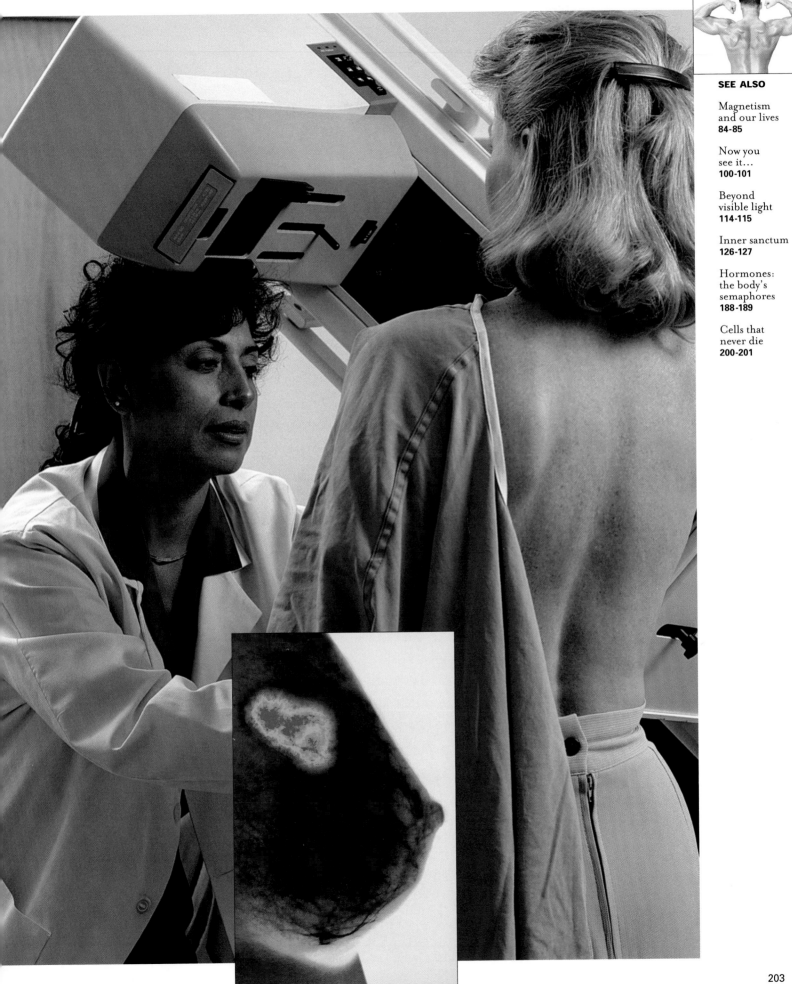

We're all Trojan horses

Consider your body: It's warm, clean, safe, and packed with food—a perfect home for all sorts of biological flotsam and jetsam that share our space.

f you ever get to feeling unattractive, consider this: Your body is so intensely appealing to trillions of individuals that they would like nothing better than to relocate there permanently. These would-be immigrants, in the form of bacteria, viruses, fungi, and parasites, are collectively responsible for infectious diseases—which is what happens when they move in and start rearranging your cellular furniture. Even when you're squeaky clean, there are as many as 20 million bacteria per square inch camping out on your skin.

Bacteria are single-celled prokaryotes—organisms that do not have separate, well-defined nuclei—and thus differ from the eukaryotic cells that make up your body. This is how chemicals called antibiotics work: They act on the cell membranes of prokaryotes but normally don't disrupt your cells.

Viruses are not cells at all. They are simply bits of genetic information—either DNA or RNA—covered by a capsule-like protein coat. They can't be destroyed by antibiotics because they're not really alive in the conventional sense. Generally you have to rely on your own biochemical defenses to rid yourself of most kinds of viruses, which is why there is no permanent cure for the common cold.

We're not completely defenseless, of course. Aside from skin, we have nasal hairs and moist membranes in both nose and mouth, including the seemingly superfluous tonsils. Damp surfaces act like microbial flypaper, trapping bacteria and exposing them to various agents. Saliva and tears, for example, contain the enzyme lysozyme and the immune-system protein immunoglobulin. Both help kill bacteria. This method is so effective that tongue infections are rare, despite a wide range of septic threats. ∎

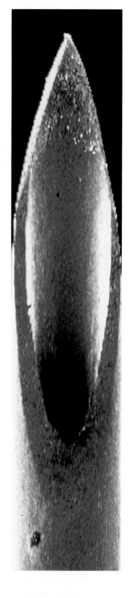

◀ **Physical barriers**
Your body's primary defense against the world's microbial hordes is physical. Skin cells reproduce so fast that the outer layer constantly sloughs off, carrying would-be residents with it. Breaching that barrier in a cut or scratch—especially with septic objects such as this hypodermic needle flecked with reddish bacteria—can be dangerous, if not fatal.

Protozoa to go ▶
Lurking inside the ravaged remains of two red blood cells they have parasitized, malaria protozoa will likely infect nearby healthy cells. The disease, rare in industrialized countries, remains a major killer in developing nations.

The time of cholera

All it takes to survive the ravages of cholera is clean water, salt, and sugar—which are often in desperately short supply in developing nations. Cholera bacteria (right) secrete a toxin that causes cells of the intestinal lining to expel sodium and water, causing intense diarrhea. Victims can lose as much as a pint of water an hour; most die of dehydration unless normal levels of fluid, glucose, and electrolytes are restored.

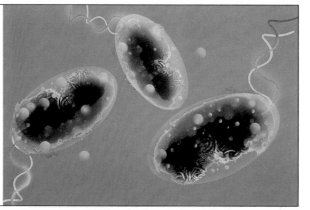

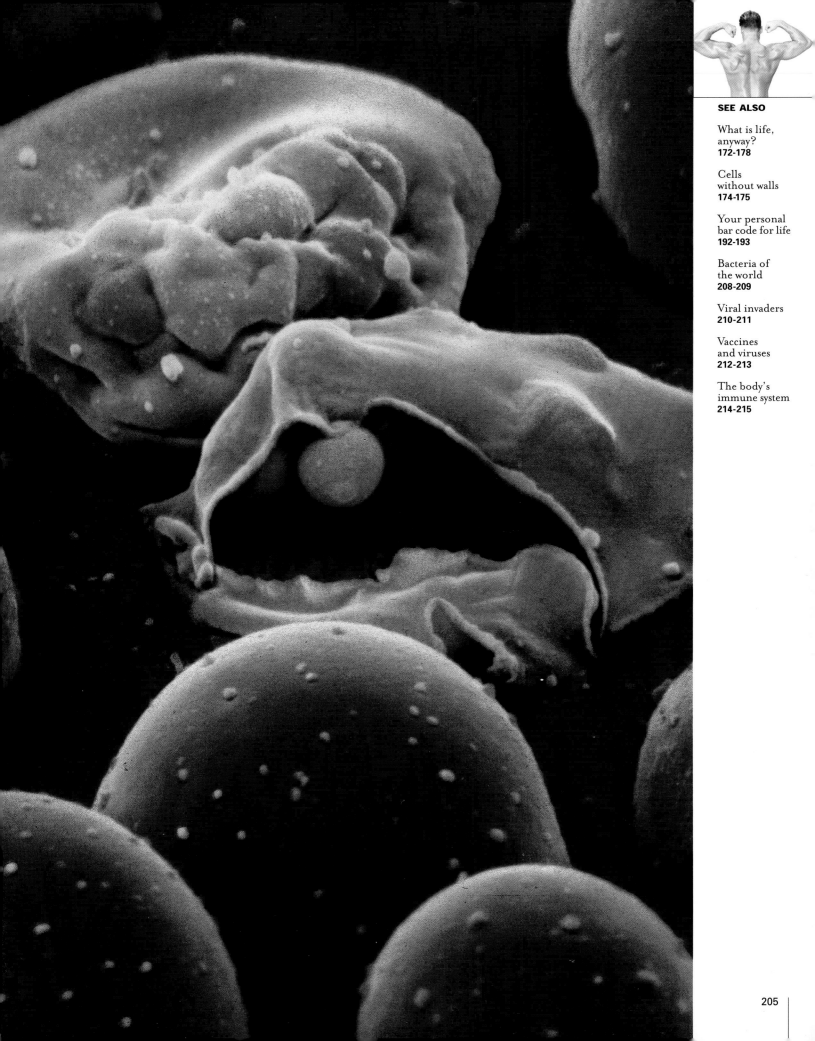

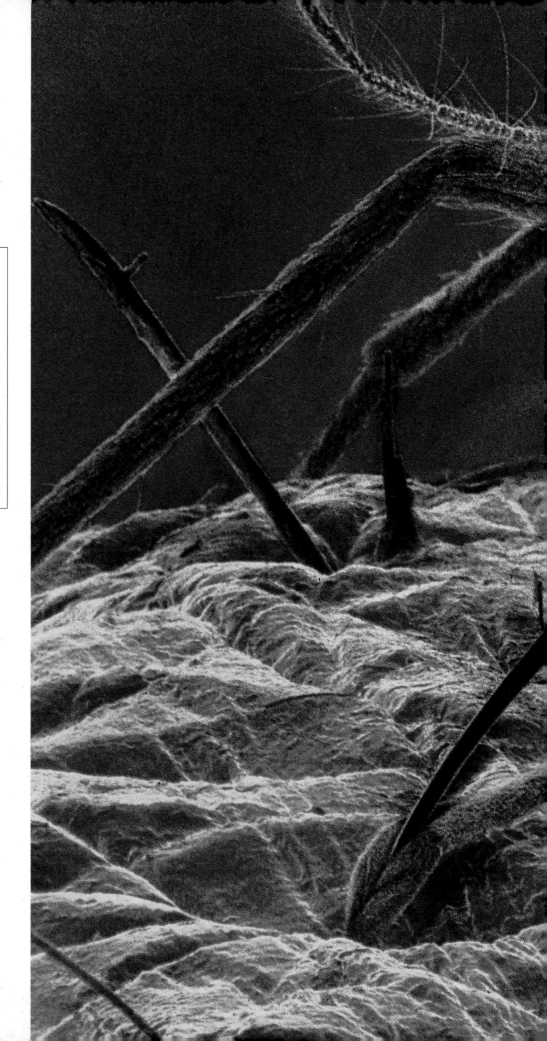

As lethal as she looks

Like a monster from another world, a female anopheles mosquito sinks her proboscis into human skin, seeking blood that will nurture her developing eggs. She also injects anticoagulant saliva—which can infect her benefactor with malaria protozoa or other dangerous hitchhikers picked up while dining on previous victims. Some 120 million cases of malaria occur every year, worldwide, primarily in tropical and subtropical countries. Another type of mosquito, *Aedes aegypti,* plays a similar role for the virus that causes yellow fever.

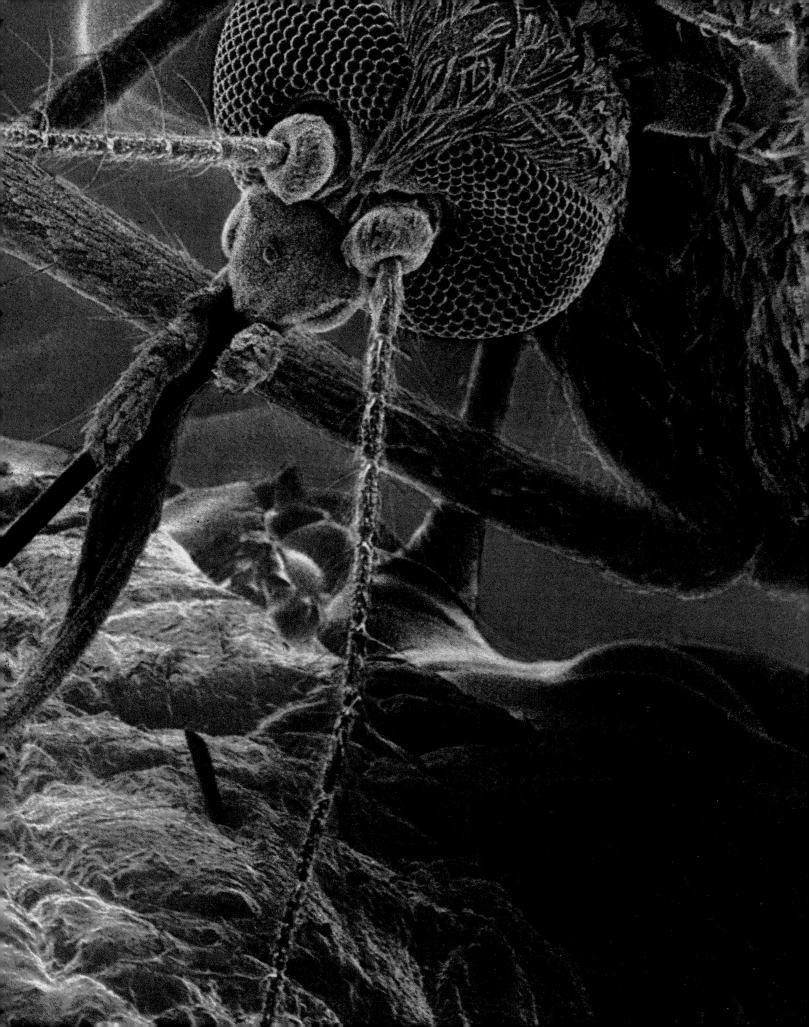

Bacteria of the world

Even the smallest scratches and nicks are chinks in our body armor,
floodgates through which bacteria can access our inner selves.

The world positively crawls with bacteria. They may be only a tenth the size of our own cells, on average, but we're covered with them. Dozens of different kinds are permanent tenants in the human gut, mouth, nose, throat, and genitals. So it's not surprising that we get sick. What's astonishing is that relatively few bacteria make us seriously ill. But those few are enough. Streptococci cause pneumonia and rheumatic fever; staphlylococci cause other sorts of maladies. Typhoid, dysentery, salmonella, cholera, and plague are caused by bacteria. So are tetanus, toxic shock syndrome, syphilis, and Lyme disease. A particularly nasty group called mycobacteria are responsible for tuberculosis and leprosy. Bacteria cause trouble in numerous ways. One problem concerns their method of penetrating human tissues, using enzymes to disable membranes and other structures that stand in their way. For example, *Helicobacter pylori* burrows into the wall of the stomach, weakening the protective mucus barrier. It's the probable culprit in nearly all peptic ulcers—and its role was only recently recognized.

Bacteria may seem simple, but they're remarkably versatile, hugely diverse, and capable of rapid evolution. There are various ways that bacteria can receive new genetic material, enabling a species to adapt to new environments or develop resistance to drugs. One is conjugation, the bacterial equivalent of sex. Another involves viruses called bacteriophages that act like hypodermic needles, injecting DNA through the membranes of cells. ■

Ready to hand ▶
There's no shortage of bacteria in modern life, as a culture of your handprint would show. In fact, the average human carries around several pounds of bacteria, the vast majority of which are not harmful. Some, such as those in the gut—home to an estimated thousand different species—perform useful functions in digestion.

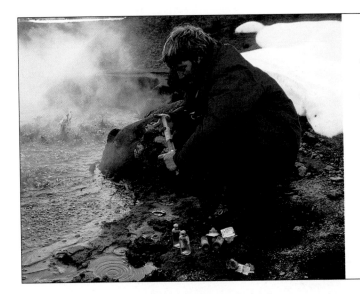

Older but wiser

There were bacteria around when the Earth was a pretty rough neighborhood, and their descendants are still with us. In the past few years, scientists have discovered bacteria that can live in temperatures as high as 250°F (left), or at the pH of battery acid, or miles below the planet's surface in solid rock, or around scalding chemical vents on the ocean floor. Some thrive on a diet of hydrogen and carbon dioxide. Most of these hardy "extremophile" critters belong to an ancient bacterial division called "archaebacteria," which was first identified in the 1970s.

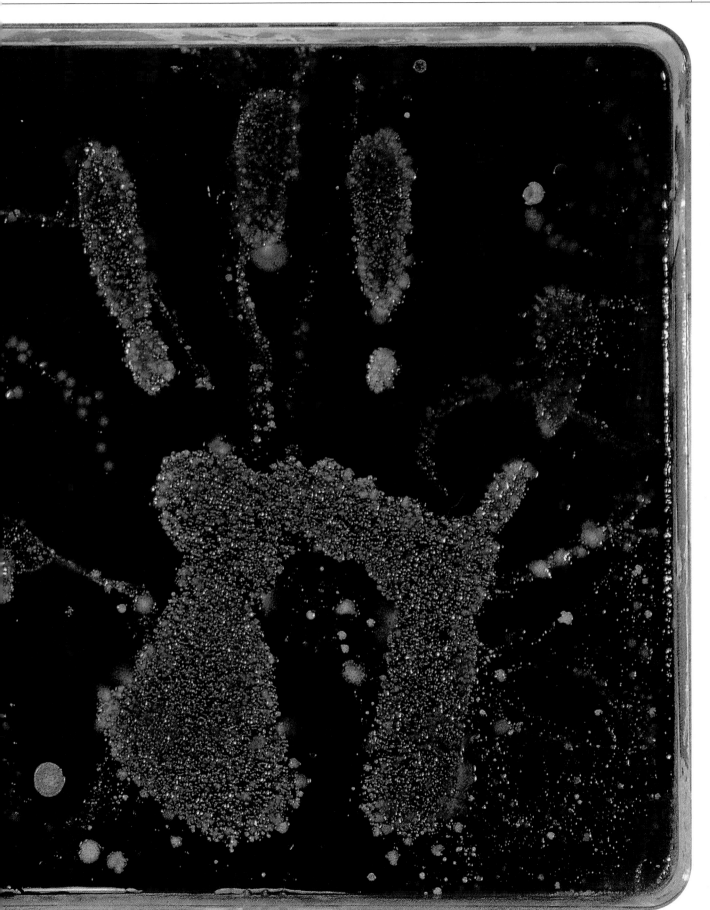

Viral invaders

In the netherworld between living and nonliving, viruses have discovered a way to reproduce—but they need help. That's where we come in.

Viruses are nothing but a set of genes on the make. As small as 20 nanometers in length, they average perhaps a hundredth the size of an average body cell—and consist merely of a few strands of nucleic acid (their total genetic material), surrounded by a simple protein coat. They cannot reproduce or make proteins by themselves. So they have to hijack your cells in order to survive—commandeering the interior machinery and nutrients, and reprogramming them to make virus parts instead of normal cell products.

To do so, a virus first must penetrate the cell membrane (below). Sometimes it does this by binding to receptors on the cell surface, which prompts the host to engulf the virus and transport it inside. Once there, the invader breaks apart its protein coat to release its genetic material. Depending on the virus, each invader may contain from a few dozen to perhaps a few hundred genes.

Viral genes either take command of various internal structures or write instructions directly into the host cell's DNA . Then the reprogrammed cell starts grabbing nearby raw materials out of its cytoplasm and manufactures viral components. When this is complete, the newly formed viruses exit their host cell—either by rupturing the membrane and killing the cell or by "budding" off, sometimes carrying a bit of the cell membrane along as a wrapper. ∎

Then and now ▶
Naturally occurring viral infections are bad enough, as the image below of Chicago police during the 1918 "Spanish flu" epidemic reminds us. But since the terrorist attacks of 2001, health officials are worrying more about the use of viruses or bacteria as weapons. Hence the protective suits and disinfectant sprays (right) used during the search for evidence of anthrax in congressional office buildings.

◀ Making them work for us
Viruses are so good at altering host cell genes that scientists want to use genetically engineered viruses to correct nature's mistakes. Some day, cystic fibrosis and other disorders may be averted by using tailored viruses to insert healthy, normal genes into people born with defective ones.

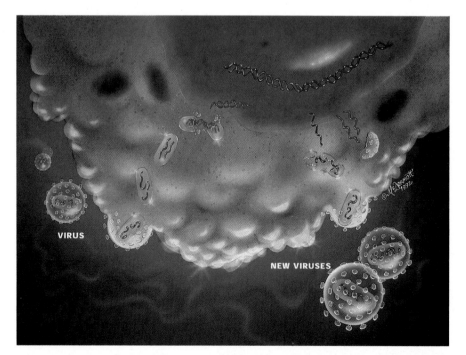

VIRUS

NEW VIRUSES

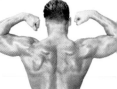

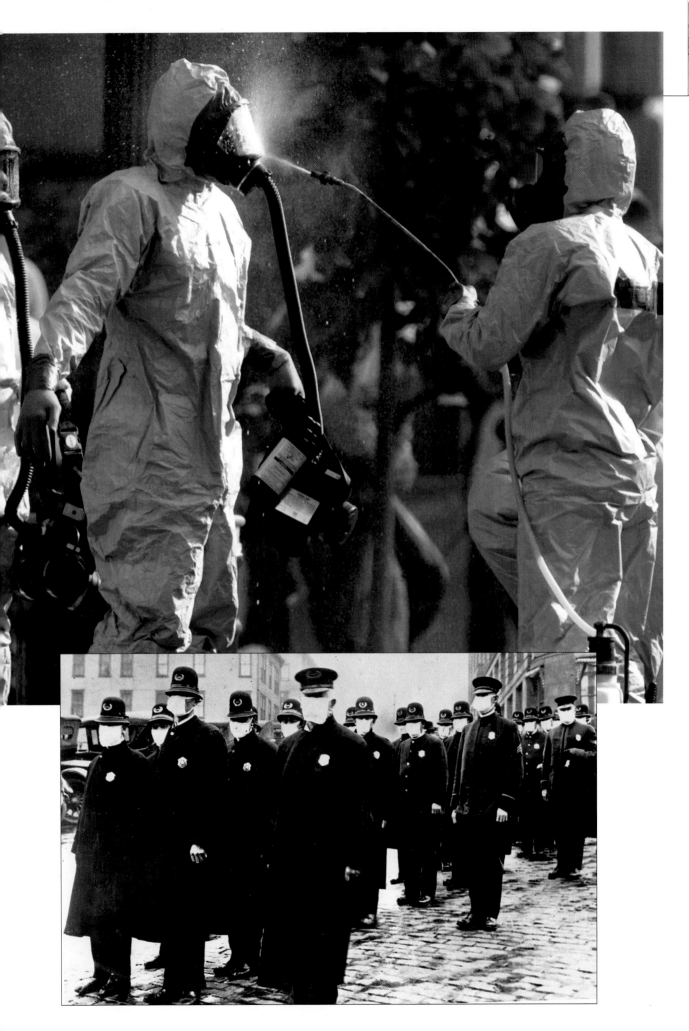

Vaccines and viruses

When natural immunity isn't enough, we can acquire artificial immunity to many diseases by vaccination. It's a real shot in the arm.

Except for children getting hypodermic injections, most of us pay little heed to one of the outright triumphs of human ingenuity: vaccination. Today, vaccines protect us from diseases that slaughtered people by the millions only a few decades ago. The concept first arose in the late 18th century, when Edward Jenner (1749-1823) discovered that inoculating a person with pus from cowpox, a cattle ailment that affects humans only mildly, would protect him from smallpox, a similar but often deadly human disease.

About 80 years later, Louis Pasteur (1822-1895) showed that injecting animals with a weakened form of a virus—recognizable by its chemistry but too damaged to cause serious trouble—also provided immunity against the real thing. In both cases, it turns out that an intelligence force of the immune system's B-cells actually memorizes the biochemical identity of the invader, as revealed by the unique compounds, called antigens, that it carries. Next time the same strain shows up, a defense is ready.

Vaccination works so well that many afflictions such as polio are on their way to becoming dim memories; smallpox already has been virtually eradicated from the planet. The last collections of smallpox virus exist in frozen cultures stored securely in a few labs, and experts wrangle periodically over whether to do away with them or not. Health officials worry that they could be used as weapons by terrorists. ∎

Mr. Super Pox ▶
Scientists date the eradication of smallpox—one of medicine's greatest triumphs—to 1978, following a massive global vaccination program. This was the first time that any disease had been totally eliminated. A celebrity of sorts, Ali Maow Maalin of Somalia became the last naturally occurring case when he contracted the disease in 1977.

The elusive flu

If only one smallpox vaccination lasts for years, why do you need a yearly flu shot? Why is it that some years, flu shots work better than others–and sometimes the reaction to the vaccine seems worse than the flu itself? The problem is the incredible variety of viral strains. Each has antigens that differ slightly—but just enough so that your B-cells can't recognize them as variants of strains they've already met. (That's also why you can get several colds a year.) So flu vaccines are concocted anew each year by combining dead samples of three virulent strains of virus that have been judged the most deserving by an international panel of researchers.

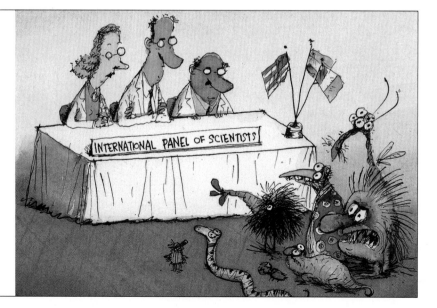

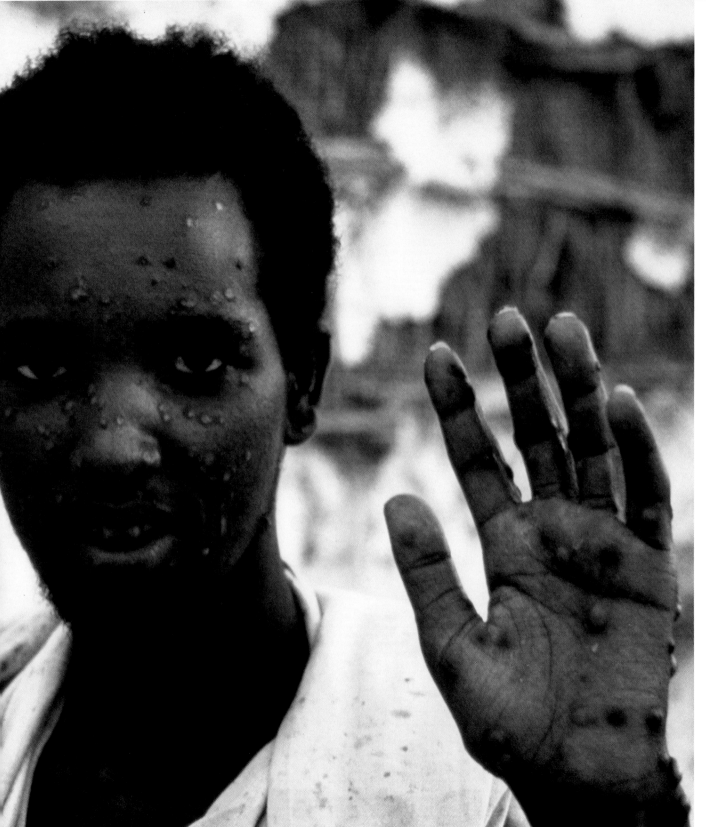

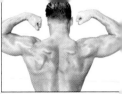

The body's immune system

With a world of microbes out to get us, how can we hope to survive?
The answer lies in one of life's most complex and amazing adaptations.

Our immune system is a powerful array of weaponry that is varied, versatile, and self-adjusting. Its level of sophistication puts the most high-tech Pentagon planner to shame. It responds to assaults in both general and specific ways.

The general stage begins when any tissue is injured. Signals from the damaged site trigger release of histamine, a chemical messenger that causes small blood vessels to widen near the injury. Various white blood cells head to the problem area, and squeeze their way through gaps that form in the enlarged walls of nearby blood vessels. Result: reddening, heating, and swelling.

Among the first to arrive are white blood cells called phagocytes, which swallow bacteria or snap up tissue wreckage and cellular debris. Some time later, your specific immune response kicks in. It has two basic divisions: an antibody recognition system operated by B-cells that handle most infections by identifying infected cells and marking them for destruction; and a cellular-attack system made up of different kinds of T-cells. One kind, "killer" T-cells, go after invaders directly by destroying the membranes of infected cells. ■

The lymph system ▶ *permeates the body like the circulatory system, and contains specialized cells to fight pathogens. These lymphocytes recognize enemies by telltale proteins or other molecules sticking out of their membranes. Those protrusions are called antigens, and every B- or T-lymphocyte has particular receptors that can bind to only one antigen. B-cells produce antibodies — protein complexes that snare antigens and neutralize or hobble them.*

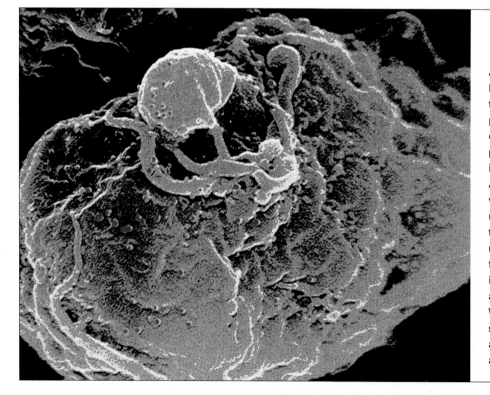

Retroviruses

Although most viruses carry their genetic information in the form of DNA, some use RNA . Within this group are the retroviruses—so-called because they produce a DNA copy from RNA, the reverse of normal cell procedures. Retroviruses pose a formidable challenge: The human immunodeficiency virus (HIV) that causes AIDS is a retrovirus. Like many other RNA viruses, HIV is more likely to mutate. But the main reason it's so deadly is that it attacks the immune system's helper T-cells, which make up about 70 percent of all T-cells and function as a kind of master control for the immune system. At left, attacking HIV appears as pink dots on the cell surface. Without helper T-cells, bodily defenses are severely impaired; cancers, other diseases, and opportunistic infections that normally are suppressed can thrive.

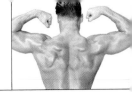

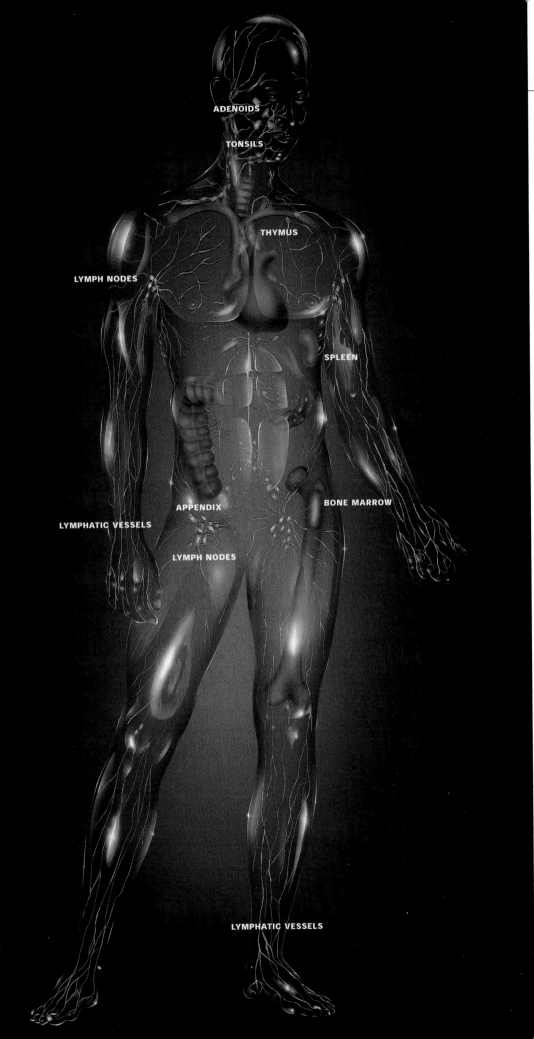

ADENOIDS

TONSILS

THYMUS

LYMPH NODES

SPLEEN

APPENDIX

LYMPHATIC VESSELS

LYMPH NODES

BONE MARROW

LYMPHATIC VESSELS

War in a lymph node

Winning the fight against infection requires constant surveillance, fast response, and an armory full of lethal, highly varied weaponry.

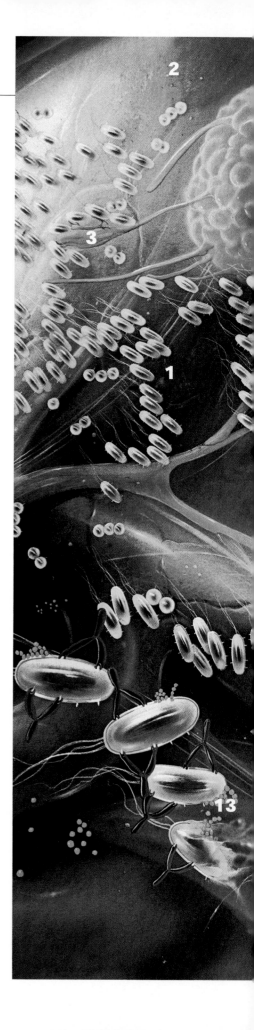

ymph nodes, a major segment of the immune system, house white blood cells and filter out microbes and other foreign particles. At times, they can be battle sites. When invading bacteria (1) pour in through a lymphatic channel (2), a macrophage ("big eating cell") (3) engulfs some invaders, digests them, and displays their unique antigen identity markers on its own surface.

This chemical message is detected by a type of white blood cell known as a helper T-cell (4) , which responds by multiplying (5) and releasing chemical messages that call more defenders to arms (6).

Meanwhile, other T-cells (7) signal B-cells (8) to join the battle and detect the particular antigen structure of the disease-causing microorganism, or pathogen (9). Some B-cells begin to reproduce (10), and these new cells store information about the antigen shape to help the body fight the same invader in the future. Other B-cells spew out thousands of antibodies (11) each second. The antibodies are Y-shaped protein structures that are as highly specialized as the antigens from the pathogens. Each binds only with one type of antigen, which it fits like a key in a lock.

Sometimes that binding process can chemically neutralize toxic effects. Alternatively, antibodies can force bacteria or other dangerous cells to clump together (12) in a process called agglutination.

But neither of these effects is as important as the antibodies' three main functions: activating a deadly chemical cascade called complement (13), which punches holes in the membranes of microorganisms; alerting and stimulating the killer T-cells that destroy infected cells directly; and binding to the exterior of bacteria or infected cells in a way that marks them for destruction.

The immune system has numerous roaming cells called phagocytes ("cell eaters") that respond to invaders whose outer surfaces have been flagged by bound antibodies. These phagocytes swallow the enemy whole. At the same time, killer T-cells grab onto marked bacteria and destroy their membranes. When this happens, the bacterial cell spills its contents and dies (14). Macrophages then clean the entire area of battle debris, engulfing scattered antibodies, dead bacteria, and other debris as the infection finally subsides. ■

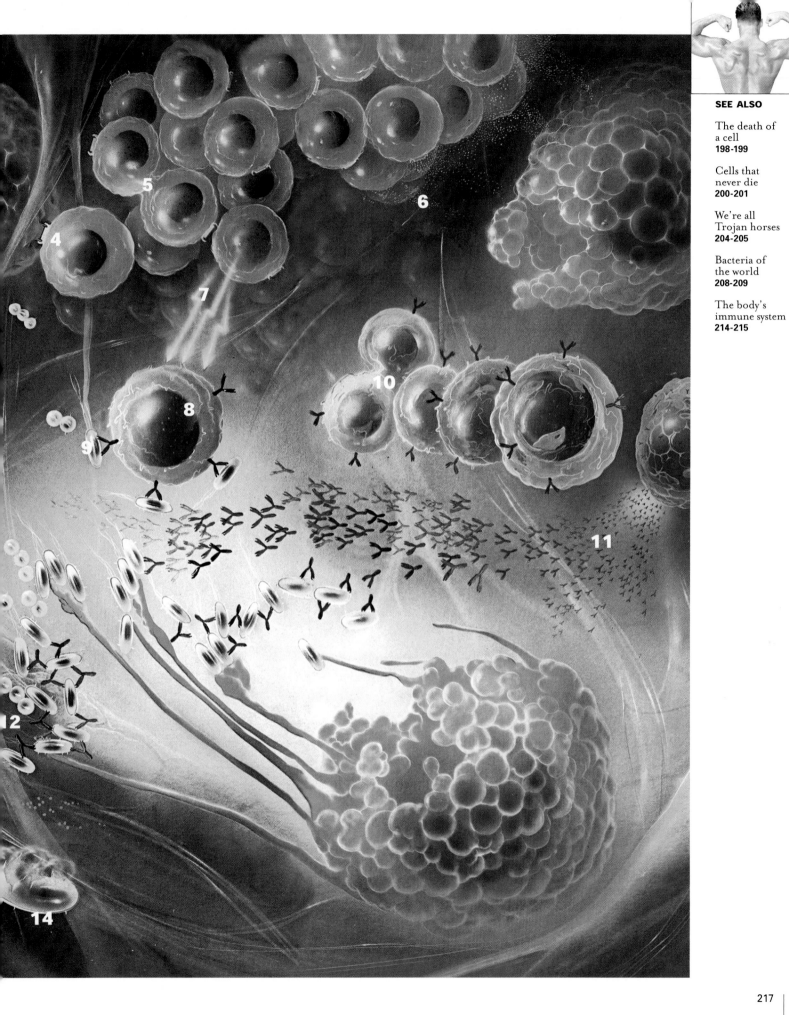

The enemy within

On the whole, our immune system works amazingly well—but sometimes it mistakes parts of ourselves for the aliens.

Anything as intricately complicated as your immune system is bound to go haywire now and again. Relatively minor errors cause the broad spectrum of complaints we call allergies. Big mistakes lead to so-called autoimmune diseases such as rheumatoid arthritis and diabetes. For some people, a single bee sting can be fatal. Their bodies respond disproportionately to danger signals, flooding themselves with histamine and other chemicals that order blood vessels to swell dramatically and blood plasma to seep out through the newly permeable capillary walls into the space between cells. Blood pressure drops catastrophically in the condition known as anaphylactic shock. At the same time, lung passages may constrict; breathing may stop.

Sometimes the immune system mistakes your own body parts for foreign tissue. Usually that's impossible, because practically every cell in your body bears your unique antigen signature on its outer surface, and that shape is recognized by the immune system's reconnaissance teams. But its general form falls into one of several groups, as in blood types. So if you ever need an organ transplant, you'll go through a tissue-typing process.

But whatever the reason, when the self-recognition process fails—or is altered by illness or exposure to an environmental agent—your body assaults its own tissue just as it would an alien microbe. The resulting damage can concern a single joint or organ—or it can affect the entire body. ■

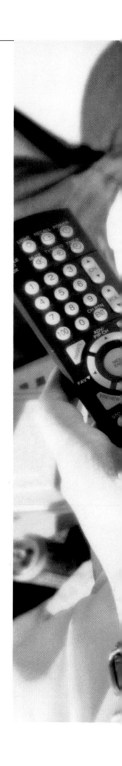

The deadly American couch potato ▶
We're killing ourselves with fat and inactivity. The United States' epidemic of obesity (which increased some 74 percent in adults between 1991 and 2003) has brought on a parallel epidemic of Type II diabetes. Type II, or "non-insulin–dependent" diabetes, frequently leads to the same kind of complications as Type I: eye, nerve, and kidney problems. Obesity is also associated with heart attacks, asthma, arthritis, and poor health in general.

Allergies: Too much of a good thing

Allergies are your immune system's way of overreacting to a trivial threat—like summoning a National Guard battalion to roust a drunk off a park bench. So it perceives ragweed pollen or dust mites or feline saliva on cat hair as a four-alarm crisis. The immune system mounts a huge counterattack complete with massive inflammation, dilation of blood vessels, increased histamine levels, and contraction of smooth-muscle tissue, particularly in the respiratory system. If the swelling is limited to the nasal area, it's called hay fever. If it involves the bronchioles of the lungs, it's asthma.

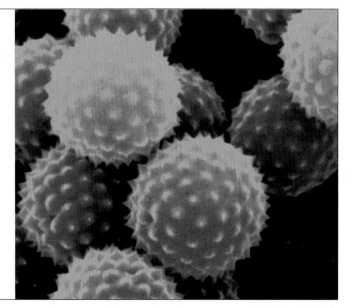

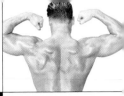

SEE ALSO

Life runs
on sugar
178-179

The body's
immune system
214-215

War in
a lymph node
216-217

Blood: the
liquid organ
242-243

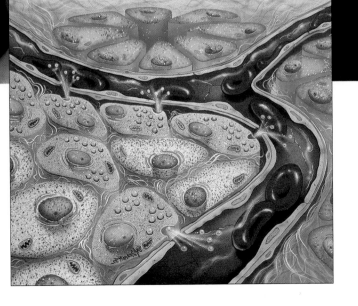

**Diabetes and the
immune system ▶**

For reasons still
unclear, T-cells at
times selectively
destroy the insulin-
making beta cells of
the pancreas (shown
as small blue spherical
emissions at right).
Without that essential
hormone, cells
cannot properly
absorb blood sugar,
and glucose builds to
dangerous levels in
the blood. That's
Type I diabetes, also
called insulin-
dependent diabetes.

Nervous systems

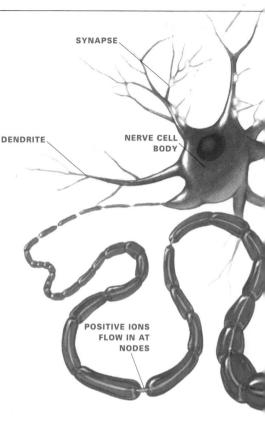

SYNAPSE

DENDRITE

NERVE CELL BODY

POSITIVE IONS FLOW IN AT NODES

Together, the brain and spinal cord—our central nervous system—are a masterpiece of signal processing and control.

Even on a good day, your nervous system runs about a thousand times slower than the cheapest computer you can buy. Yet it is incomparably more powerful—and more subtle—thanks to the elaborate electrochemical method of communication that nerve cells have evolved to network signals back and forth. The nerves in your body range in length from less than one millimeter to more than a meter, but they all talk to each other in the same general way.

Most neurons, or nerve cells, have a central body, a set of branching filaments called dendrites that connect to hundreds or thousands of other cells, and a long, insulated transmission fiber called an axon. Neurons almost never touch; at the point of connection, they are separated by a tiny gap, or synapse, about a millionth of an inch wide. Before that point, a nerve cell transports its signal down the axon as an electrical charge. At the synapse, the message becomes chemical. The receiving neuron then responds, sending yet another electrical message down the line. ∎

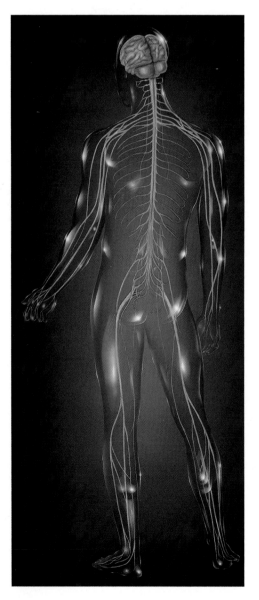

◀ **Hello, central**

The central nervous system processes incoming sensory information and initiates responses such as ordering muscle motion. It communicates with the rest of the body via the fibers of the extensive peripheral nervous system.

1. Incoming information

Each neuron can receive different chemical signals from hundreds or thousands of other neurons. Each signal can either excite or inhibit the receiving cell. If the net sum of all incoming messages tells the neuron to fire, it undergoes a chemical change that lets positive ions flow in from outside and build up an internal electrical charge.

2. Neural currents

The positive electrical spike propagates down the neuron's long axon until it comes to a synapse, a gap leading to another neuron. Here the electrical signal provokes small vesicles to move toward the gap and disgorge their load of chemicals.

RECEPTOR

RECEIVING NEURON

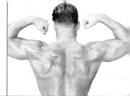

INSULATING SHEATH

AXON

SYNAPTIC GAP

FIRING VESICLE RELEASES NEUROTRANSMITTER

INTACT VESICLE

3. Bridging the gap
*Vesicles release
neurotransmitters
that bind to specific
receptors on the
receiving side.
Nearly a hundred
neurotransmitters
are known, giving
your central nervous
system tremendous
versatility.*

4. Push "reset"
*After each event,
neurotransmitters
have to be cleared out
of the synapses,
usually by enzymes
or reabsorption by
the sending cell.
Drugs such as Prozac
inhibit that process.*

The control center

Next time you do something really stupid, don't blame your equipment. Nature has entrusted us with the most complicated entity in existence—the human brain, shown here in a series of CAT scans. This three or so pounds of tapioca-like goop (which consumes an astonishing 20 percent of your blood sugar at rest!) holds around a hundred billion neurons, and many of them can each interconnect with 10,000 or more others. That means that the number of different paths an impulse can take through the brain is unimaginably huge, approaching if not actually exceeding the number of particles in the known universe. The fact that some of us can barely balance our checkbooks is not for want of adequate tools.

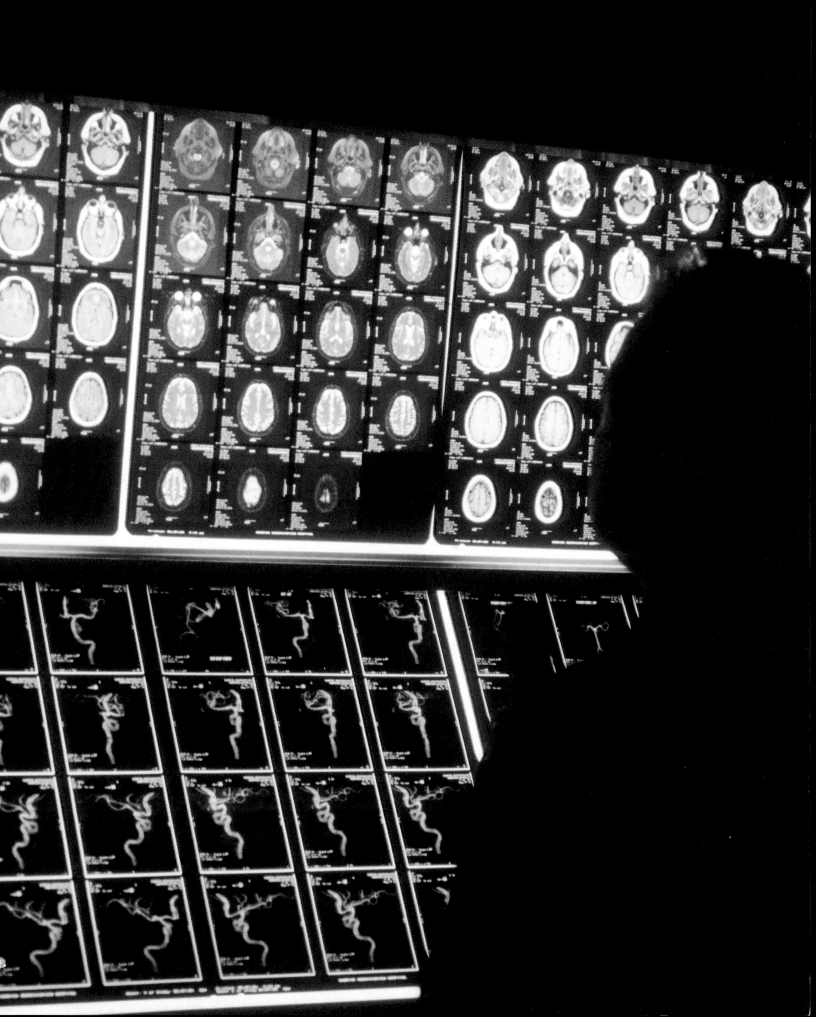

Nerves and muscles

Even the slightest move you make is living proof of teamwork: Muscles and nerves act together because they can't do it alone.

When you turned to this page, you exercised the most familiar part of your nervous system, voluntary motor operation. Unlike the autonomic nervous system, which automatically takes care of housekeeping functions such as blood pressure and heartbeat, motor neurons work only on specific orders. They're attached to muscle fibers by a junction that resembles a synapse, though it works somewhat differently because the signals are always excitatory.

Motor nerves terminate in a buttonlike blob separated by a narrow gap from the muscle cell's end plate, which is covered with receptors sensitive to a neurotransmitter called acetylcholine (ACh). When the nerve fires, a blast of ACh moves across the gap, immediately binding to the receptors. The muscle opens its ion channels, develops an electric charge, and contracts down its length. Once that job is done, you have to get rid of ACh in a hurry. As long as it's there, the muscle will remain rigid and voluntary control will be impossible. So the end plate is equipped with an enzyme, acetylcholinesterase (ACh-ase), that inactivates ACh after each contraction.

Many popular insecticides work by inhibiting the action of ACh-ase. In effect, they leave insects' muscles locked into maximum tension, resulting in paralysis and death. In very large exposures, they can also be toxic to humans. ■

Getting on ▶
Some effects of aging take place in the efficiency of our neuromuscular junctions. With advancing years, one theory goes, part of the acetylcholine released by an arriving nerve impulse leaks away, diluting the signal. Animal studies indicate that an increasing number of acetylcholine receptors on the muscle cells of older subjects experience reduced effectiveness. C'est la vie.

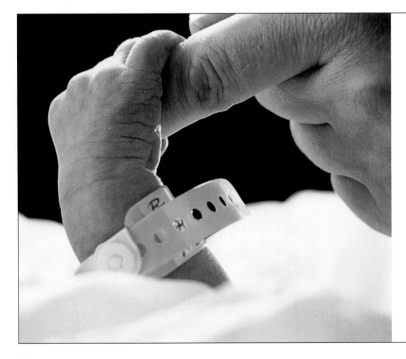

Wired: the reflex arc

Often our muscles react before we're really aware of it. Those are reflexes: rapid, involuntary reactions that are the same each time for a given stimulus. That's because each reaction bypasses the cerebral cortex and is controlled at the upper spinal cord, speedily transmitted over circuits known as reflex arcs. In each case, a sensory impulse—the stimulus—speeds along a sensory nerve fiber to a cell body in the spinal cord. The response goes straight back to the muscle, skipping the brain. Because it involves circuitry that's far simpler than those ferrying messages to the brain, this type of reaction often takes place before the same sensory impulse reaches the cortex and is consciously interpreted as pain.

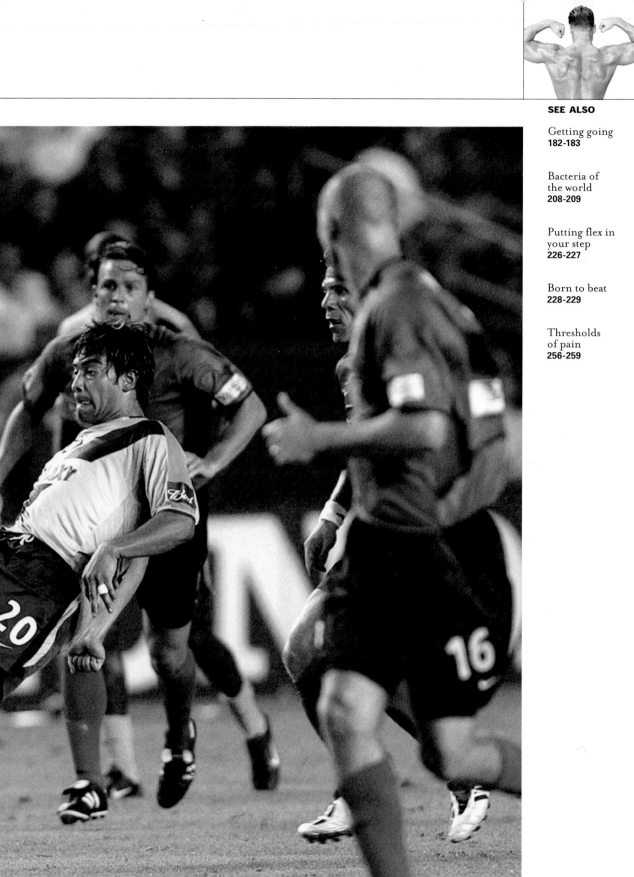

Putting flex in your step

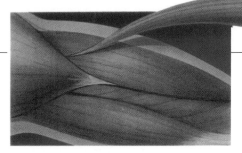

Like interlocking jigsaw pieces, two interactive proteins make possible all our bodily motions—to say nothing of spectacular abs, pecs, and delts.

You're constantly using your muscles, even when you're not moving. Just standing upright requires constant tension in tens of thousands of strands. How do muscles get that kind of pull? The answer lies in the interplay of two kinds of proteins with ATP and positively charged calcium ions. Although there are four different types of muscle—skeletal, cardiac, smooth, and myoepithelial (the ones that squeeze milk, saliva, or sweat out of tissues)—they all contract in the same general way. Each muscle fiber (actually a single cell that has many nuclei because it formed from the fusion of multiple cells) is a long, cylindrical structure made up of two slightly overlapping protein filaments. One is myosin, a thick bundle out of which protrude numerous "cross-bridges" that look rather like the heads of golf clubs. The other is actin, a thinner, more streamlined filament. When a muscle contracts, myosin tugs on actin like hands pulling in a rope, and the strands slide past one another.

Those painful twinges we call cramps stem from repeated and immoderate demands on muscle tissue. They can result from lactic acid buildup due to anaerobic work ("the burn"), from sweating out too much sodium and potassium (both essential to muscle chemistry), often brought on by strenuous exercise at high temperatures. Changing position or eliminating stress usually helps relieve cramps. ■

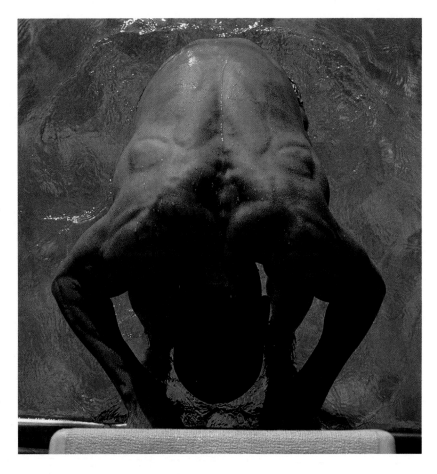

◄ Ready to move
There are two basic kinds of muscle contraction: isotonic and isometric. In the former, muscle fibers visibly move and the tension remains constant as the muscle changes length. In the latter, shown by this swimmer, the muscle tissue contracts, but the muscle is prevented from changing length. Tension keeps increasing, but the actin-myosin complexes just spin their wheels. Either kind of exercise adds to muscle strength.

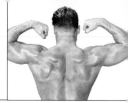

MUSCLE FIBER

MYOFIBRIL

CONTRACTILE UNIT

When actin meets myosin ▶

Each myosin "head" has a special receptacle for ATP molecules, which create a chemical tension so that the head tends to curl and snap to the side. This energy can be released only when the myosin head contacts an actin molecule—but that doesn't happen in resting muscles, because a molecule of troponin lies between the actin and the myosin.

RELAXATION

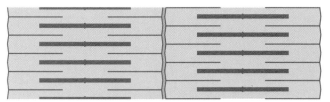

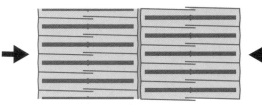

CONTRACTION

When a muscle receives orders to contract, it releases calcium ions that bond to the troponin, pulling it out of the way. Myosin hits actin; pent-up energy is released as the myosin head snaps sideways and pulls the actin strand along— all in a few thousandths of a second. Each time more calcium is released, the muscle fiber tugs again and the actin slides a bit.

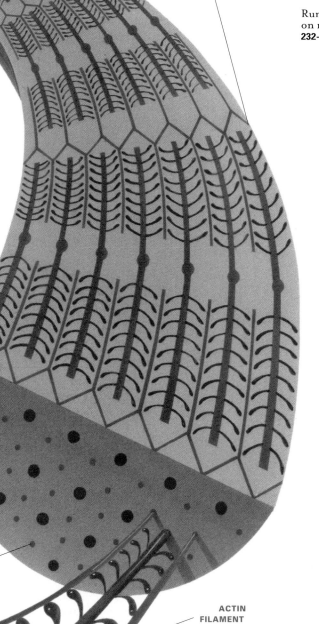

MYOSIN

ACTIN

MYOSIN HEAD WITH ATP

FLIPPED HEAD RELEASING SPENT ATP

ACTIN FILAMENT

MYOSIN FILAMENT

Born to beat

Various civilizations have prized it as the seat of emotion, even of knowledge or the human soul. We see it more simply, as a tireless and near-perfect pump.

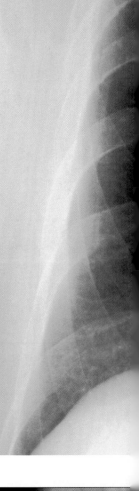

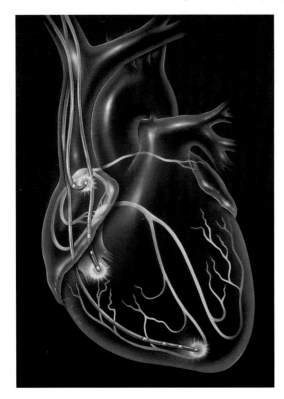

f your heart could talk, it wouldn't have time to say anything. It's far too busy beating 100,000 times every day, and doing it hard enough to shove all five quarts of blood once around your body every minute. It's actually two pumps: One atrium and ventricle serve just your lungs, while the other two chambers handle everything else. It began pulsing around four weeks after conception—when your entire embryo was only about the size of this O. At birth, it automatically reconfigured its own plumbing, to route blood to lungs it didn't use until that moment. Since then it's just been pumping away, billions of times.

Unlike skeletal muscle, heart muscle never tires. Every heart muscle cell is welded to its neighbors by tough protein-fiber connections and joined by electrical junctions that allow voltages to pass easily. Thus, contraction in one cell triggers contraction in the next and the next, creating waves whose rhythm is governed by a master pacemaker. This little blob of cells, called the sinoatrial node, sits at the top of the right atrium. Its electrical timing signal spreads from the top of the heart to the bottom, slowing down a bit as it passes through a less conductive layer between the atria and the ventricles. As a result, muscles in both atria contract first, followed an instant later by the ventricles, which squeeze from the bottom up. Lub-dub.

At rest, about 27 percent of your cardiac output goes to the digestive tract, and 20 percent to the kidneys. Skeletal muscles get 15 percent, the brain about 13 percent, the skin 9 percent, and the bones 5 percent. Only 3 percent flows to the heart itself. But even moderate exercise changes everything. The heart triples its requirements, skeletal muscles take 64 percent, and the flow for digestion and kidneys drops by at least half. Your skin, to shed more heat, now takes nearly 14 percent. ■

◄ **The beater**
Sometimes heart cells can get out of synchronization. They start to beat chaotically in a random pattern called fibrillation. This process can often be rectified by blasting the heart muscles with a large electrical charge that suddenly stops all the action like a judge banging a gavel in a courtroom. In the ensuing calm, the muscle cells begin following the pacemaker again. Defibrillator devices are increasingly found in offices and public areas.

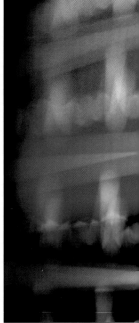

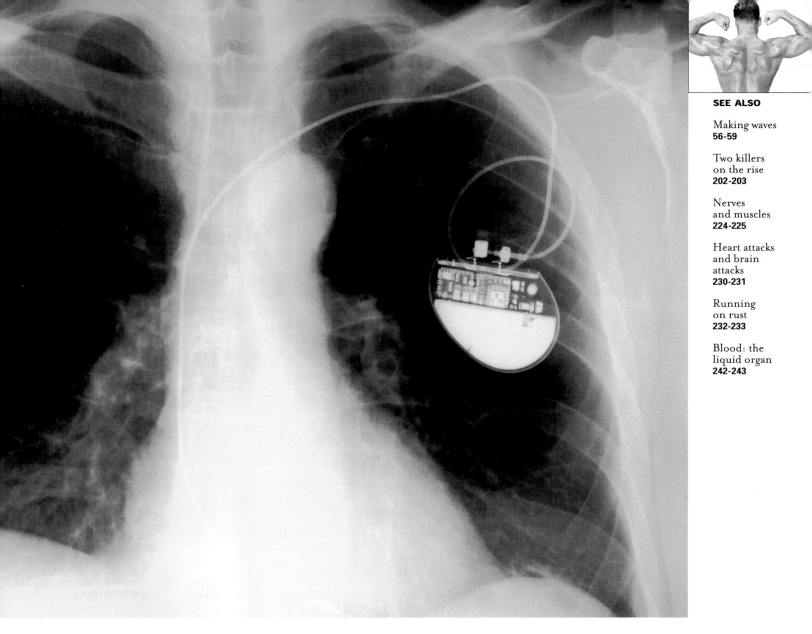

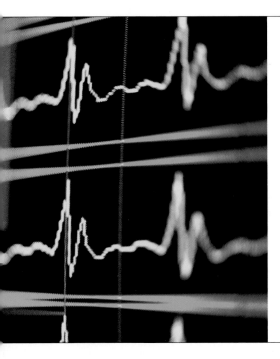

Keeping up the pressure

As it navigates through some 60,000 miles of blood vessels that supply the human body, our blood squeezes through up to 40 billion capillaries. Capillaries are only about seven microns wide—so small that red blood cells must pass through in single file. Every time your ventricles contract against that resistance, your blood pressure rises. Normally, the pressure of contraction—systole—would support a column of mercury about 5 inches (120 mm) high; relaxation pressure—diastole—normally supports a little over 3 inches (80 mm), giving the clinically "normal" reading of "120 over 80." The pressure at any given spot, however, shifts constantly.

▲ **Setting the pace**
If the main pacemaker fails, the heart takes its cadence from either of two other natural pacemakers: One in the lower right atrium, and another at the bottom of the ventricles (far left). Or you can get an artificial pacemaker (above) that kicks in as needed if the natural system is defective.

Heart attacks and brain attacks

We have the same problem with our two most valued organs. It's simple and alarmingly widespread: Constricted arteries can leave both of them gasping.

t seems ridiculous to imagine the heart not getting enough blood. Yet its impermeable lining prevents heart muscle from taking advantage of the nutritious fluid that constantly flows through it. Instead, its cells must rely on the coronary arteries that wrap around the outside of the heart on both sides. It is essential that the flow in these vessels remains high. Heart muscle works so hard that it consumes 65 percent of the oxygen carried to it, compared to approximately 25 percent in less demanding tissues. Cardiac function can't continue very long on anaerobic metabolism; the heart is the body's aerobic tissue par excellence. Moreover, its muscles can't be replaced. Exercise can make them bigger, but not more numerous. By the time you were one year old, you possessed all the heart muscle cells you were ever going to get. Protecting that investment is the single most important thing you can do to prolong your life, since heart disease is the nation's leading cause of death (29 percent), followed by cancer (23 percent) and then stroke (7 percent). Even if you survive a heart attack, lots of cardiac muscle may die. ■

Hardworking brain

Even though you can never remember Aunt Edith's birthday, your brain is a metabolic dynamo. While you're reading this, your three-pound brain—only a couple percent of your total body weight—is getting about one-sixth of your heart's total output and is consuming one-fifth of all the oxygen in your blood. Unfortunately, cranial arteries are highly vulnerable to small clots and blockages. Neurons won't last more than two or three minutes without oxygen. Once they die, the parts of the body they control are paralyzed, temporarily or permanently. Because of the way the brain is wired, damage to the left side causes paralysis on the right side of the body, and vice versa.

Clogged pipes ▶
Fatty deposits called atheromas build up inside an artery wall, restricting blood flow and encouraging clots. Once an already narrowed artery gets plugged, the muscle cells served by that artery simply die of oxygen and glucose starvation. Scar tissue forms in their place. Annually, about 1.1 million Americans have a heart attack. About 44 percent die—nearly half of them in the first hour after onset.

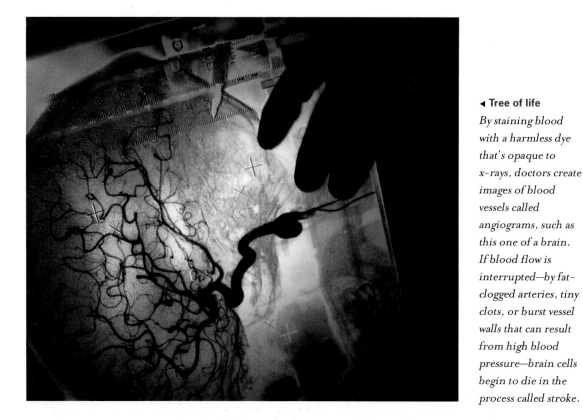

◀ Tree of life
By staining blood with a harmless dye that's opaque to x-rays, doctors create images of blood vessels called angiograms, such as this one of a brain. If blood flow is interrupted—by fat-clogged arteries, tiny clots, or burst vessel walls that can result from high blood pressure—brain cells begin to die in the process called stroke.

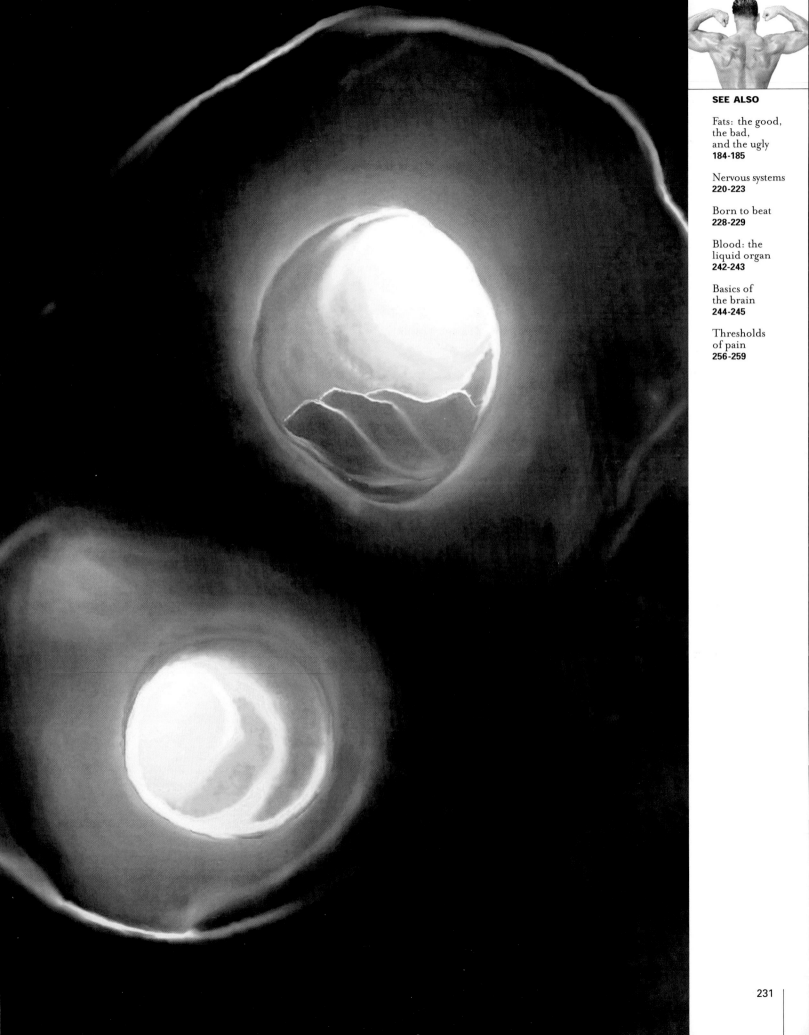

Running on rust

Using blood to move oxygen around the body can be tricky, because plasma can't carry enough. That's where hemoglobin comes in.

Aerobic creatures like us need a lot of oxygen. When we breathe, a sophisticated gas-exchange system enables our blood to take in oxygen and give off carbon dioxide—a waste product of aerobic metabolism. The key is hemoglobin, an iron-rich compound that has the versatility to bond either with oxygen or carbon dioxide—and to release either molecule, as conditions change. If the iron in our blood floated freely, it would bond tightly to the oxygen, just as it does in the outside world, as rust.

In a resting state, our bodies need about half a quart of pure oxygen every minute. We get that easily, for even quiet breathing brings in perhaps 8 quarts of air a minute, or 1½ quarts of pure oxygen. Strenuous exercise can increase oxygen demand 20-fold, so our lungs have a much greater maximum capacity: About 300 quarts per minute in males, and about 250 quarts in females. There's lots of room, because pulmonary tissue is so densely convoluted. By one estimate, if the gas-exchange area of your lungs were spread out flat, it would cover a tennis court. ■

Oxygen, energy, and blood

A lab-bound athlete makes the moves that affect his rate of oxygen consumption, as well as other physiological processes. Physical exercise pumps up the body's need to consume more oxygen and expel more carbon dioxide. Exactly what specific physiological signal triggers these increases is unclear. But it appears that as the brain signals skeletal muscles to move, it also turns up the dial on respiration—even before those muscles get going. As exercise proceeds, the cranked-up cardiovascular system delivers more oxygen. To determine the body's peak effort in laboratory studies, the subject's workout is allowed to proceed to the point of exhaustion, and breath samples are collected for analysis of oxygen and carbon dioxide levels. Each liter of consumed oxygen translates into almost five kilocalories of energy used by the body.

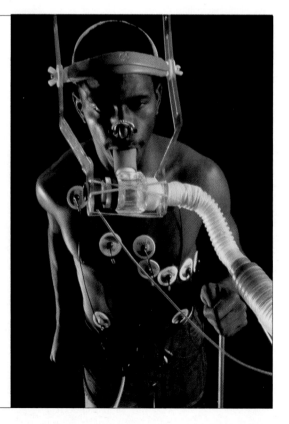

Lung function ▼
Oxygen enters your blood through 700 million alveoli—tiny air sacs surrounded by even smaller capillaries only one cell wide. It's easy for gas exchange to occur on that scale. Only about 1.5 percent of blood oxygen is dissolved in liquid. The rest bonds temporarily to iron-rich hemoglobin molecules in red blood cells. Interestingly, carbon monoxide binds so strongly to hemoglobin that it prevents it from carrying oxygen.

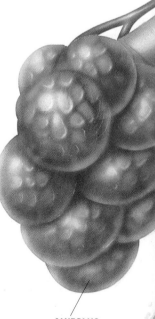

ALVEOLUS

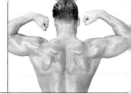

Adaptation ▶

Research into the physiology of sherpas—as well as other groups who are superbly adapted to the thin air at high altitudes—has revealed that they do not breathe faster. Instead, they appear to have a larger "tidal volume" of lung space, with more and larger alveoli.

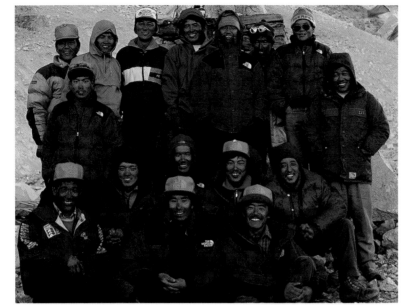

SEE ALSO

It's all downhill from here
10-11

Ashes to ashes, rust to rust
158-159

What is life, anyway?
172-173

Turning on your body heat
180-181

Getting going
182-183

Putting flex in your step
226-227

Blood: the liquid organ
242-243

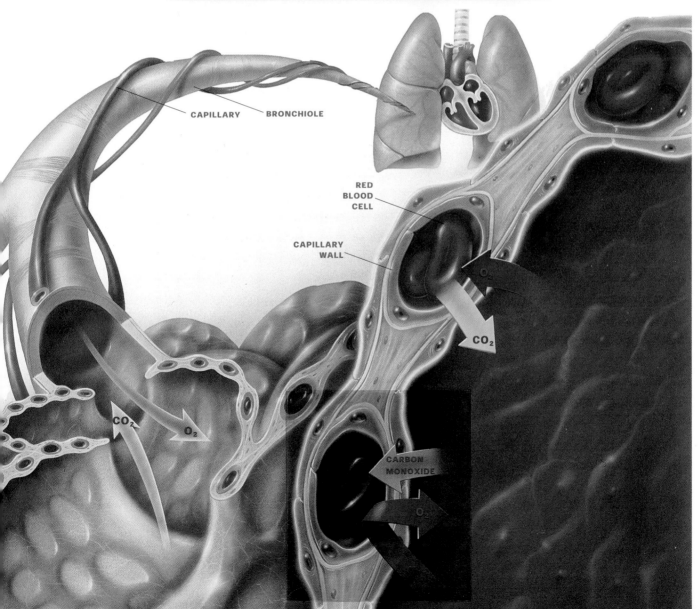

CAPILLARY   BRONCHIOLE

RED BLOOD CELL

CAPILLARY WALL

O_2

CO_2

CO_2 O_2

CARBON MONOXIDE O_2

Being good to your gut

Next time you feel utterly stuffed after a meal, content yourself with this thought: Much of what you just ate will never enter your body.

Because your 30-foot-long digestive tract is basically a sealed tube with portals only to the outside world, its contents really aren't part of you at all. Sure, much of the nutritional content of the food you eat will be absorbed across various membranes. But digestion is basically an outside job. Good thing, too, since most of your cells could never tolerate the abuse your dinner takes—the corrosive bath of gastric juices (more acidic than battery acid), the flesh-dissolving enzymes, or the host of bacteria lurking in your colon.

As you chomp on a slice of pizza, saliva soaks it with antibacterial agents and amylase, an enzyme that starts the breakdown of starches. Often, just the thought of eating is enough to spark salivation, priming the digestive processes.

Then the wad passes into the stomach, a ten-inch-long food processor with a trap door at each end. Every day, your stomach secretes as much as two or three quarts of gastric juices, mucus, hydrochloric acid, and an enzyme called pepsin. That deadly mix kills still more bacteria, continues breaking down carbohydrates, and begins the process of cracking complex proteins. No nutrients are taken up; only aspirin and alcohol are absorbed directly through the stomach lining—although most of both are processed in the small intestine—which is why that drink "goes right to the head."

After pummeling your pizza into mush, the stomach contracts in wavelike motions, squeezing the food bit by bit into the labyrinthine small intestine. Here, sodium bicarbonate from the pancreas sloshes in, neutralizing the acid. Some protein-busting enzymes also enter, while detergentlike bile made in the liver and stored in the gall bladder emulsifies fats.

Rhythmic contractions shove the goo along. Most carbohydrates break into simple sugars; proteins crack into amino acids. Fats are converted into fatty acids and glycerol. By the time your chow reaches the colon, there's nothing left of it but very complex carbohydrates such as cellulose, a lot of microbes, and some water. The bacteria go to work, manufacturing some vitamins, which are absorbed into the intestinal membrane. Finally, your pizza is history. ■

Hole in the wall?

Once thought to be caused by stress and over-acidity, stomach ulcers result almost entirely from bacterial infection by *Helicobacter pylori,* a microbe that is amazingly well equipped to survive the organ's highly corrosive secretions. Once the stomach wall is breached, gastric juices attack, enlarging the hole. Bile salts from the small intestine also may enter the stomach, exacerbating any ulcers and causing symptoms of heartburn, stomach pain, and nausea.

Sponging up ▶
Carbohydrate digestion begins in the mouth and stomach. But in terms of final absorption, the small intestine is where the action is. In each of its three main parts, partly digested food is further broken down by various enzymes added earlier. This organ, in 21 feet of length, boasts a surface area about ten times that of the skin. Microvilli (below) snag amino acids, fatty acids, glucose, vitamins, salts, and glycerol.

Gut feeling ▼
Millions of villi—tiny, fingerlike projections of the inner intestinal lining—are further divided into thousands of microvilli. This remarkable lining is replaced every three days or so.

MICROVILLUS

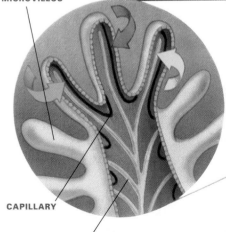

CAPILLARY

LYMPH VESSEL

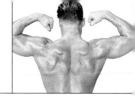

SEE ALSO

Seeking
balance: pH
150-151

The building
blocks of life
186-187

Bacteria of
the world
208-209

The magic
of the liver
236-237

Vitamins
and minerals
238-239

Bones and teeth
240-241

▼ **Digestion:**
an all-day affair?

*Your digestive system
takes a day and a half
at most to process
food, mechanically
and chemically, from
start to finish—maybe
half a ton per year.*

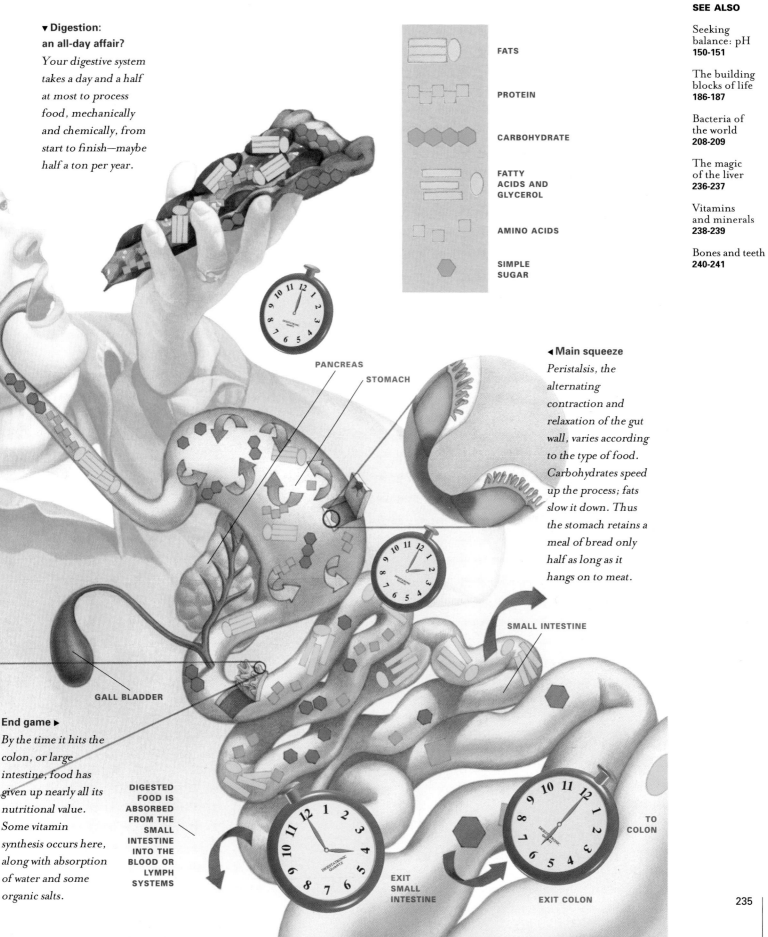

FATS

PROTEIN

CARBOHYDRATE

FATTY
ACIDS AND
GLYCEROL

AMINO ACIDS

SIMPLE
SUGAR

PANCREAS

STOMACH

◄ **Main squeeze**

*Peristalsis, the
alternating
contraction and
relaxation of the gut
wall, varies according
to the type of food.
Carbohydrates speed
up the process; fats
slow it down. Thus
the stomach retains a
meal of bread only
half as long as it
hangs on to meat.*

SMALL INTESTINE

GALL BLADDER

End game ►

*By the time it hits the
colon, or large
intestine, food has
given up nearly all its
nutritional value.
Some vitamin
synthesis occurs here,
along with absorption
of water and some
organic salts.*

DIGESTED
FOOD IS
ABSORBED
FROM THE
SMALL
INTESTINE
INTO THE
BLOOD OR
LYMPH
SYSTEMS

EXIT
SMALL
INTESTINE

EXIT COLON

TO
COLON

235

The magic of the liver

It digests, it detoxifies, it gobbles up dead and injured cells. It also manages to store excess sugar. In fact, it's your most important metabolic organ.

ou'd think that up to a full day of complicated chemical processing would be sufficient to turn the constituents of your dinner into ooze you can use. But in fact the job isn't done until everything passes through the liver. So essential is this three-pound, multifunctional masterpiece that nature has provided you with about eight times more of it than you absolutely need, in case some gets damaged by injury, disease, or martinis.

Uniquely, the liver receives blood from two sources: The hepatic artery supplies oxygen, while the portal vein brings nutrients straight from the digestive tract. Inside the liver, blood is routed through tiny structures called lobules, where single cells carry out the organ's 500 or so different functions. These duties fall into three basic categories: conversion, storage, and excretion. Among other tasks, the liver creates blood proteins from amino acids, detoxifies toxins, hangs onto fat-soluble vitamins, helps regulate blood sugar, and dumps bilirubin, a breakdown product of red blood cells. Then it ships everything out via the hepatic vein. ∎

Dealing with alcohol

Of the many things liver cells detoxify, ethyl alcohol poses a particular challenge: Its energy content is fairly high, less than that of fat but more than that of carbohydrate. But your body has no way to store it. So the liver uses it preferentially as an energy source, instead of the fatty acids it normally employs, and tucks those unused fats away in surrounding tissue. That's why heavy drinkers tend to have enlarged livers.

◄**Chemical microprocessors**
The liver contains tens of thousands of polygonal lobules—each made up of hundreds of hepatocytes, or liver cells, served by arteries, veins, and bile ducts. Together, hepatocytes perform the organ's myriad metabolic functions. They convert raw nutrients into essential substances and make poisons benign.

Where everything has its place ►
The liver accomplishes its wide variety of tracking, processing, and storage with almost no cell specialization. Just about any cell can make fats out of proteins and sugars, or vice versa.

LOBULE (SIMPLIFIED)

BILE DUCTULE

HEPATIC ARTERY (BRANCH)

PORTAL VEIN (BRANCH)

HEPATOCYTE

BILE DUCTULE (BRANCH)

SEE ALSO

Life runs
on sugar
178-179

Fats: the good,
the bad,
and the ugly
184-185

The building
blocks of life
186-187

Being good
to your gut
234-235

Vitamins
and minerals
238-239

Blood: the
liquid organ
242-243

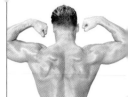

Many jobs to do ▶

Liver cells store substances the body will need in the future, such as glycogen, a starchlike form of glucose. They also send various compounds into the bloodstream—both manufactured products for general distribution to various body cells and waste material for elimination. In addition, they make bile, a detergentlike fluid needed for digesting fats, from worn-out red blood cells. Bile travels along ductules to the gall bladder for temporary storage.

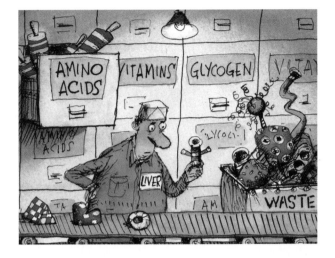

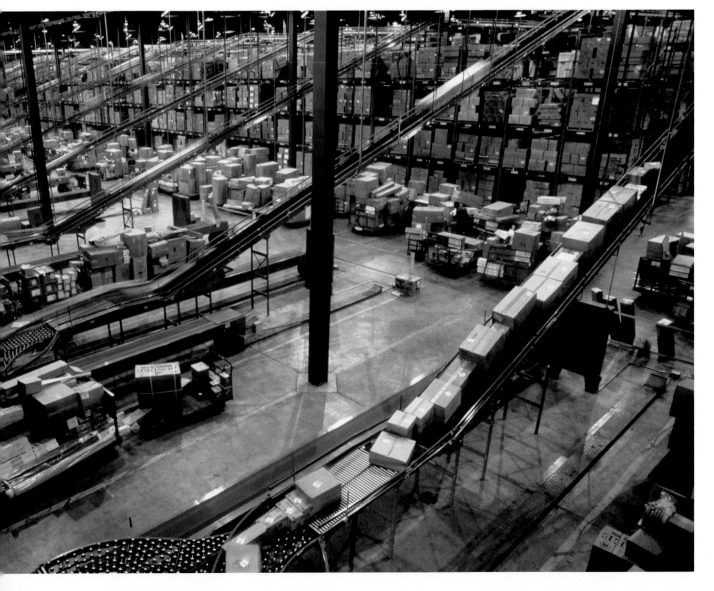

Vitamins and minerals

All of us need all of them; they're absolutely essential to a healthy life.
But a little goes a long way—and you can get too much of a good thing.

Somewhere along the evolutionary line, the human body seems to have lost its ability to manufacture certain substances so integral to life that we call them vitamins. You need about 20, as well as a number of essential minerals, to conduct your complex biochemical business. The only one you create by yourself is vitamin D, which is produced by the action of sunshine on the skin. Many Americans get plenty of vitamins from the foods they eat. Yet thanks to the poor dietary habits of the world's most embarrassingly overfed nation, most physicians now recommend that Americans take a multivitamin every day, since four-fifth of us don't eat enough fruits and vegetables.

That's particularly important in the case of the water-soluble C and B-complex vitamins: They're consumed rapidly on a daily basis, and any excess intakes are regularly flushed out in urine. But the fat-soluble vitamins—A, D, E, and K—remain stored in the liver or other fatty tissues for weeks or even months. In particular, you may need to take special vitamin supplements if you're pregnant, taking certain antibiotics, on kidney dialysis, or either elderly, anemic, alcoholic—or if you just don't eat well.

Vitamin C has a host of uses. It helps key enzymes produce healthy teeth, bones, gums, and connective tissues, as well as some neurotransmitters and hormones. It also aids the immune reaction, which is why some burns, injuries, and infections draw down body levels of this vitamin. But like all vitamins, it's effective in relatively small amounts. Many people take huge doses in the belief that they will somehow prevent colds or mitigate the symptoms; there is no conclusive evidence for this notion. In fact, taking more than a gram of vitamin C a day can produce diarrhea, kidney problems, and stomach distress. As a rule, you're unlikely to be deficient in this vitamin unless you're an alcoholic or a

What's so great about broccoli?

Oxygen is extremely reactive; it exists in your body in several hazardous forms, including free radicals—pieces of broken compounds desperately seeking new chemical bonds. Some can cause mutations in DNA, encouraging cancer. Your body contains substances such as vitamins C and E that bind with or otherwise incapacitate oxygen radicals, and are hence called antioxidants. They are abundant in cruciform vegetables such as broccoli and cauliflower.

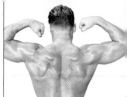

◀ **Clean your plate**
*Researchers stress
that a healthy daily
diet should consist
of: 6-11 servings of
bread, cereal, rice, or
pasta; 2-4 servings of
fruit; 3-5 servings of
vegetables; 2-3
servings of meat,
poultry, fish, dry
beans, eggs or nuts;
2-3 servings of milk,
yogurt, or cheese;
and a bit of fat, oil,
and sweets.
We don't do it.*

smoker. Carbon monoxide, a principal component of auto exhaust and cigarette smoke, depletes vitamin C.

Vitamin B_{12} and the seven-member B-complex function chiefly as coenzymes. That is, they help enzymes ensure that a desired chemical reaction happens. The B vitamins are essential to a number of metabolic processes cells use to obtain energy and break down carbohydrates and fats. In addition, folic acid and B_{12} are necessary for maintenance of DNA and RNA. Folic acid also plays a crucial role in fetal development of a normal nervous system; deficiencies can cause severe birth defects. Other B vitamins have critical roles in maintaining skin tone, bone chemistry, nerve function, digestion, and sleep rhythms. In general, surplus intake of B_{12} and B-complex do not harm the body.

These fat-soluble vitamins ensure proper cell function, red blood cell activity, vision, and protection from certain toxins. Vitamin A makes up part of the light-sensitive, purple pigment called rhodopsin that is used in the eye's rod cells, essential to night vision. It is necessary for proper skin tone—which is why similar substances, called retinoids, are used to treat damaged skin. It also helps maintain resistance to some pathogens. In addition, there is tentative evidence that vitamin A and its precursor, beta-carotene, may have a protective effect against certain cancers—although some studies have raised serious doubts about this claim. Vitamin E prevents anemia, and may help reduce the risk of heart disease.

Vitamin D aids in maintaining proper absorption of calcium and phosphorus in the body. Very little is needed, but the elderly especially should make sure they get enough. Vitamin K acts in the liver, affecting the manufacture of substances that make blood clot. You get it from green, leafy vegetables and through the action of intestinal bacteria.

But even that's not enough, no more than flour and water are enough to make a cake. In addition to vitamins, you need seven inorganic elements—calcium, sodium, chlorine, phosphorus, potassium, magnesium, and iron—as well as trace amounts of a dozen others, including such apparent oddities as selenium, zinc, cobalt, and chromium. The thyroid hormone thyroxin, which controls the rate of glucose metabolism, can't be made without a small amount of iodine. Before corner drugstores and iodized salt, the largest source in most diets was seafood. Those who lived in regions where ocean fish didn't exist and soils contained little iodine were prone to develop thyroid swellings called goiters, still common in developing countries. Calcium, sodium, and potassium, needed for neuromuscular function, serve as electrolytes. Bone can't be built without potassium and calcium. You're more closely related to rocks than you may prefer to think. ∎

Bones and teeth

They may not be as lively as some other tissues, but these mineral bastions of our bodies are a lot more dynamic than they seem.

You might think that once your bone and joints are fully formed, they don't change unless you break a bone. In fact, bone constantly refurbishes itself to meet new demands, adding calcium-bearing minerals to—or removing them from—each bone's framelike template. That template is made of collagen, a rubbery protein that enables your skeletal components to bend slightly without breaking. The filler, hydroxyapatite, consists largely of calcium bound to phosphate groups; it makes your skeleton nearly as strong as concrete, pound for pound.

Bone-making cells, called osteoblasts, wriggle into this matrix and secrete minerals, entombing themselves in their own creations. Though they no longer move, they live on, sustained by small blood vessels in channels that make bone porous. Meanwhile, other cells called osteoclasts constantly tear down the structure, either to make room for replacement material or to scavenge calcium that the body needs elsewhere. Bone is especially responsive to forces of gravitation and stress: The harder you push on it, the thicker it gets. That's why the right arm of a right-handed tennis player may have 30 percent more bone than the left arm. That's also why "weightless" astronauts lose bone mass during prolonged orbiting. ■

Taking a bigger bite ▶

Like bones, teeth are largely made of hydroxyapatite, both the inner dentin and the shiny outer enamel. Humans are equipped with two sets: 20 "baby" or deciduous teeth, and the replacement set of 32, which push up (x-ray, right) to replace the deciduous set. A child's mouth simply isn't large enough to accommodate 32 teeth. So the permanent ones arrive only when the jaw is large enough.

◀ Inside a broken bone

Repairs begin almost as soon as a break occurs. A blood clot, rich in debris-cleaning phagocytes and osteoclasts, forms at once. Cartilage creates a callus to replace the clot and hold the break together. Ruptures in the connective tissue stimulate production of osteoblasts, which work with osteoclasts to create new bone at the site.

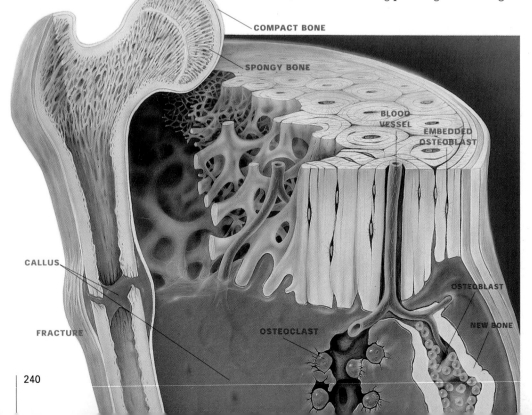

COMPACT BONE

SPONGY BONE

BLOOD VESSEL

EMBEDDED OSTEOBLAST

CALLUS

FRACTURE

OSTEOCLAST

OSTEOBLAST

NEW BONE

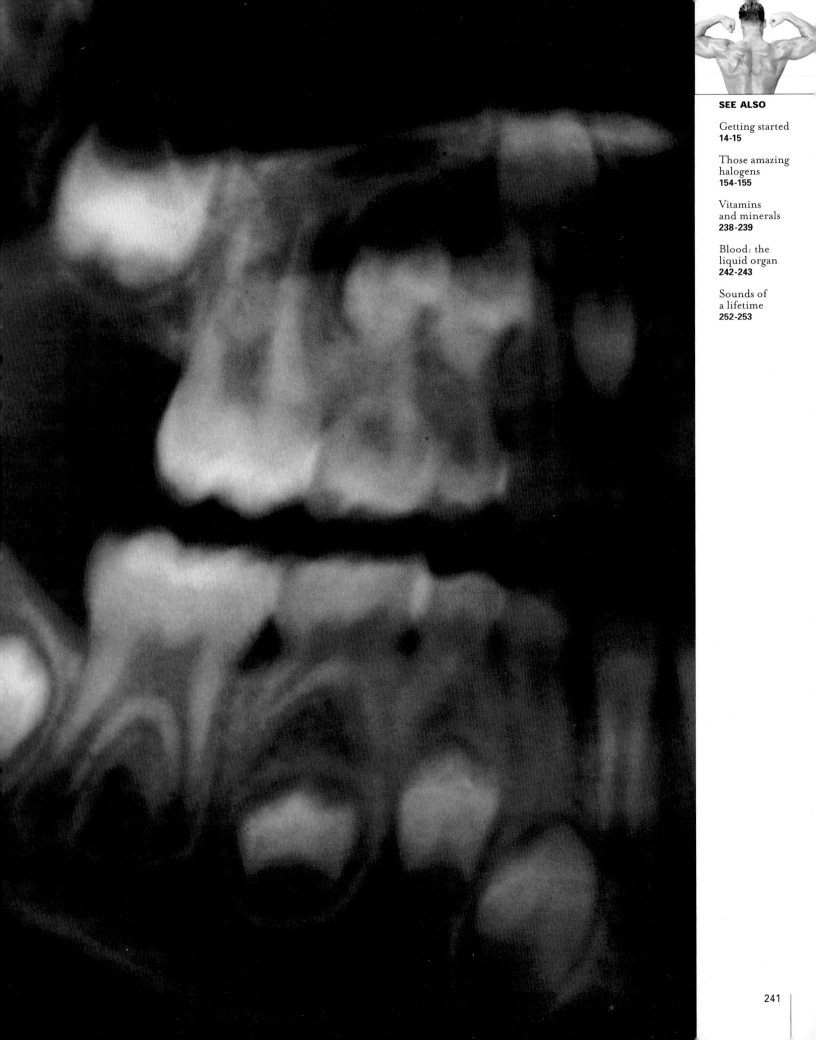

Blood: the liquid organ

A crimson river flows through you, coursing some 60,000 miles as it twists past nearly every cell and structure, nourishing and cleansing and regulating.

Blood actually is thicker than water—for excellent reasons. Normally, every milliliter of this life-sustaining liquid harbors some five billion red blood cells, seven million white blood cells, and 250 million platelets. The first, rich in hemoglobin, ferry oxygen throughout the body; the second comprise five general types of immune-system cells; the last are involved in blood vessel repair. Together, they make up about 45 percent of whole blood. The rest is the watery plasma that carries proteins, sugars, fats, electrolytes, minerals, and more.

Each of us contains around five liters of blood, some 8 percent of our total body weight, on average. Arterial blood—rich in oxygen because it has recently passed through the lungs and is now on its way to the rest of the body—is bright red in color. Venous blood, returning to the heart after having exchanged significant amounts of oxygen for carbon dioxide, has a darker tinge. Gases, nutrients, and other chemicals usually enter or leave the bloodstream through the tiniest blood vessels, called capillaries. Their extremely thin walls facilitate such exchange. Only seven or so microns wide, capillaries hold about 5 percent of the body's blood supply at any given time, but the total area available for chemical diffusion is huge: larger than a basketball court.

Blood is a motley mixture. If your doctor orders a CBC, or complete blood count, the lab looks for the relative abundance of each cell type. Disk-shaped red blood cells really are red; their color stems from the iron-containing portion of hemoglobin, which turns ruddy when it combines with oxygen. You've got 25 or 30 trillion of them at any time, and you're replacing them at the rate of about three million a second. The relatively tiny platelets clump together to make blood clot. Unusually numerous T-cells may be indicative of an active immune system. CBCs usually include a breakdown of the various kinds of white cells, or leukocytes, present. Although they normally make up less than one percent of blood by volume, leukocytes can be a valuable indicator of infection if their types and their proportions are known. Blood tests can also reveal concentrations of specific antibodies, thus identifying particular infections. ∎

Blood cholesterol levels

Lipids aren't soluble in water, so they're escorted through the bloodstream by two kinds of water-soluble combinations called lipoproteins. Low-density lipoprotein (LDL) is the "bad" cholesterol that can help fats accumulate on the inner walls of arteries. High-density lipoprotein (HDL) is the "good" form that tows fats toward the liver for processing. High HDL is associated with lower incidence of artery disease. Total cholesterol (LDL + HDL) shouldn't go too far over 200 milligrams per tenth of a liter of blood.

Blood type ▶
Our red cells determine basic blood type, which is a function of whether either of two different antigens—called A and B—exist on the surface of those cells. Your immune system attacks whatever blood cells display an antigen pattern that is different from yours. Type O blood contains neither antigen; type AB has both. Thus, people with type O blood— about half of the population—cannot safely accept donated blood of any other type—but their blood can be used safely by all people.

A single team of specialists ▶
Awash in the clear, straw-colored liquid called plasma are the blood's various cells: red, white, and platelets (inset). Each is manufactured by a different mechanism and has a specific set of tasks to perform.

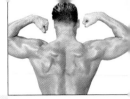

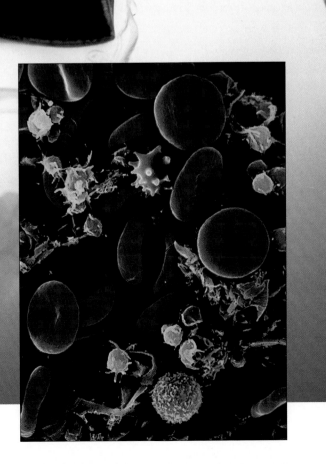

Basics of the brain

Like a house that has undergone extensive additions over the years, the brain is made up of four distinct parts, each slightly younger in evolutionary terms.

The four main parts of the human brain are stacked atop one another in chronological order. The two lowest and oldest layers, the brain stem and cerebellum, handle basic functions such as breathing, eye movement, balance, circulation, and swallowing. Above them is a cluster called the limbic system, associated with primal emotions and instinctual responses such as fight or flight. It affects the initial processing of memory.

On top of everything else is the relatively new cerebrum. Separated into two crenellated hemispheres, it mediates conscious thought, stores memory associations, interprets language and vision, oversees sensations from your sensory apparatus, and directs motor control for your voluntary actions.

Each hemisphere controls the opposite side of the body. Although each is a virtual mirror image of the other, there is considerable specialization: The left side is much more involved in language and logical operations; the right in spatial relations, music, and pattern recognition. Both halves are interconnected by some 300 million axons. Indeed, every part of the brain ultimately is wired into every other part, through a wondrous neural assemblage of loops that feed forward as well as back. As a result, there's no such thing as an "entirely rational thought" or a "pure emotion." Highly intellectual thoughts are colored by raw feelings and memory, and even the most primitive of impulses is always modified, to some extent, by consciousness. ■

Cells that don't think ▶
The neurons in your brain account for only about half of its bulk. The rest is taken up by glial cells that handle maintenance. Mere blood is not allowed here; brain cells are nourished by their own special fluid containing only those molecules that can get through a highly selective filtration system known as the blood-brain barrier.

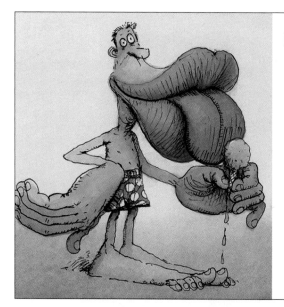

If we looked the way we feel

Had all the different parts of our body been sized according to how many sensory nerves they contained, we'd wind up looking something like this. Lips and fingertips have far more nerve endings—and thus far greater sensitivity—than, say, elbows or buttocks.

The brain's sensory cortex is organized accordingly. Located along the upper rear of the cerebrum, it processes incoming information from various sense organs. Major portions are devoted to interpreting messages from relatively small but important body parts, such as the ears and eyes.

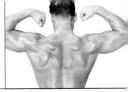

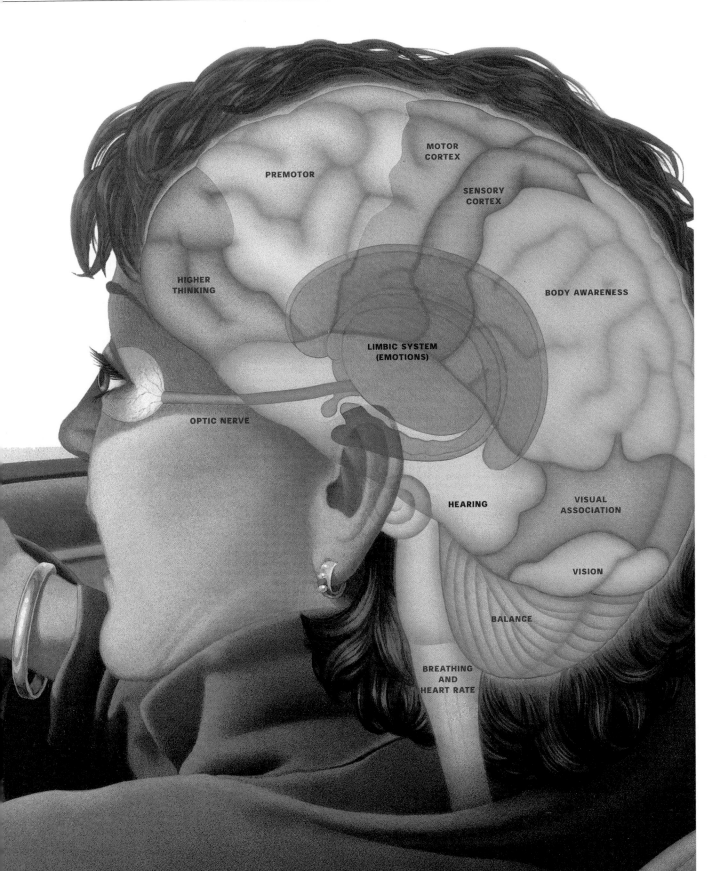

MOTOR
CORTEX

PREMOTOR

SENSORY
CORTEX

HIGHER
THINKING

BODY AWARENESS

LIMBIC SYSTEM
(EMOTIONS)

OPTIC NERVE

HEARING

VISUAL
ASSOCIATION

VISION

BALANCE

BREATHING
AND
HEART RATE

Seeing and believing

The eyes sense light and color, transforming the energy of photons into electrochemical messages. But it is the brain that interprets—and perceives.

Contrary to the popular saying, what you see is definitely not what you get, at least on the basic anatomical level. Much of what we "see" is an artifact of image processing by the optical system.

For example, if what we perceived were a true representation of reality, we'd all see a big black hole near the center of the visual field. That's because there's a sizeable blind spot on every retina, a place with no photoreceptor cells at all because the space is taken up by the optic nerve connection. Why don't we see these blind spots all the time? Because the brain just "paints in" the missing area of each eye by using visual data from the other eye.

The human eyeball contains about a hundred million rods, complex, light-sensitive nerve cells that see only in shades of gray but respond well to low-light situations. It also has about three million color-perceiving cones, which are most sensitive to bright light. Cones come in three different types, each favoring one of three ranges of wavelengths: red, blue, or green.

At the back of the retina, almost exactly opposite the pupil, lies a small spot called the macula. Its very center contains a tiny depression known as the fovea. Composed of densely packed cones but no rods, the fovea is the seat of greatest visual acuity. But with age, the macula often undergoes progressive breakdown: macular degeneration. Blood vessels behind the retina can leak, causing profound loss of central vision. Sometimes, laser procedures can seal off the leakage; more often, nothing can be done. ■

Visual teamwork ▶
When light hits the retina (far right), it encounters the light-sensitive rod and cone cells (right). Photons prompt pigments within the rods and cones to split, sparking a series of chemical steps that ultimately result in a signal being sent to the optic nerves.

Rays and pupils ▶
The optic nerves from each eye extend separately a couple of inches, then come together at the optic chiasm. There, information from the right half of each eye's visual field is joined, then routed to the left side of the brain; left-side images from both eyes go to the right. The visual cortex analyzes vertical vs. horizontal, light vs. dark, background vs. foreground, and so forth, before combining them all into what we perceive as reality.

Improving your reception

If you could miniaturize yourself and get inside a human eye, you'd notice that the light rays that fall on the retina—the eye's innermost, photosensitive layer—create an image that is upside-down relative to reality, because the lens in each eye inverts incoming light. That upside-down image wouldn't be as sharp as you might expect, either, because the retina enhances the contrast between darker and lighter areas before it moves a picture along to the brain for processing. It's amazing that we can see straight.

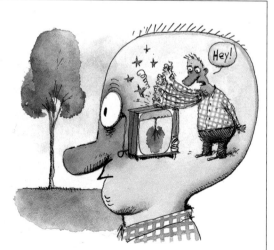

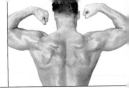

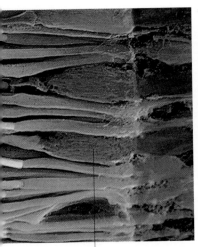

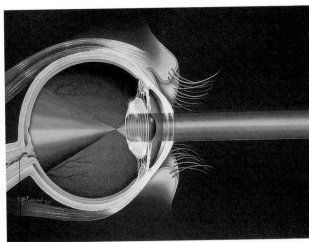

CONE

TO OPTIC NERVE

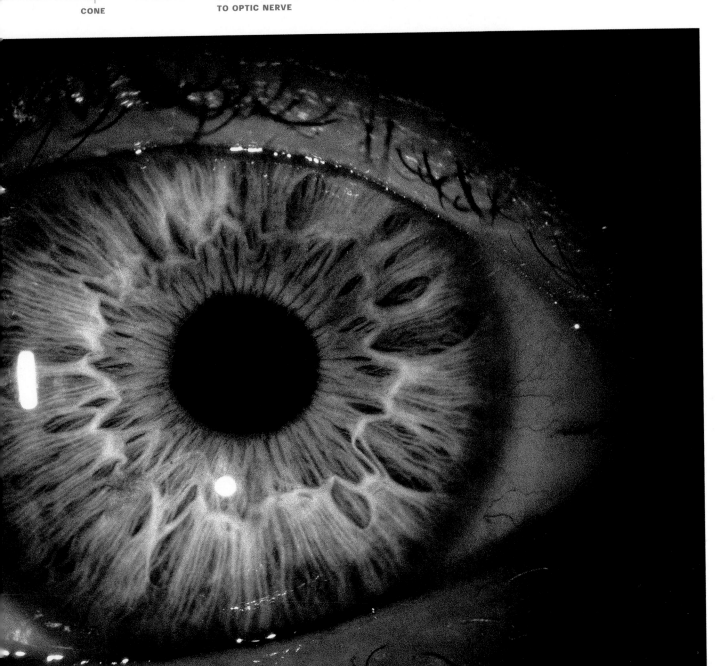

Perception—or confusion?

Do we see what's there, or just what we want to see? Like life itself,
it's a mix of experience and innate nature.

Vision is all in our heads. Much of what we consider seeing is actually highly processed perception, and the human brain has hundreds of procedural short-cuts and darkroom tricks to speed up this task. In particular, it is splendidly adept at recognizing objects and patterns.

But those same talents, along with the fact that what we "see" is often a function of visual context, make it easy to fool the eye. Many times we perceive patterns that aren't there, or interpret information incorrectly as a by-product of the way that our brains process visual information. The examples on these pages represent only a tiny fraction of such phenomena.

On the other hand, the brain is also incredibly good at dealing with incomplete or ambiguous visual imagery and making split-second "best-guess" decisions. For example, a small child can recognize his grandmother at the train station—even though it's dark and he sees only her silhouette or perhaps a mere fraction of her face. This ability to fill in a situation from a few scraps of information—which we all employ many times every day—no doubt helped our earliest ancestors avoid being eaten by predators. ■

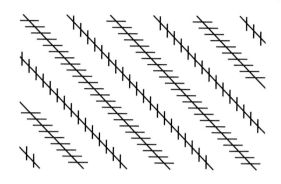

▲ Parallel or not?
Though they seem out of alignment, all nine of these lines are perfectly parallel. You might expect a simple explanation for something so dramatic. Obviously, cross-hatching somehow plays a role. But scientists are still struggling to understand exactly what's up with this misperception. One theory, angular displacement, refers to the fact that we usually overestimate acute angles.

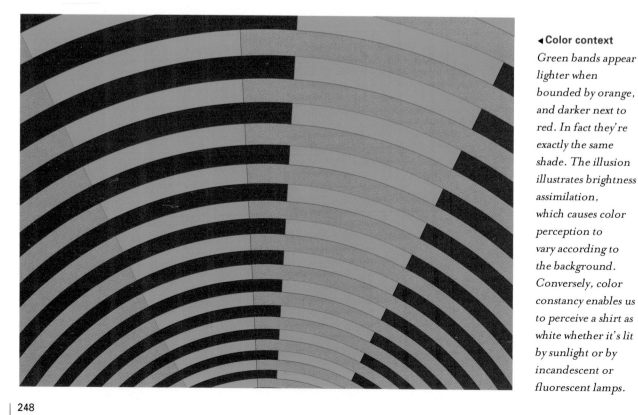

◄ Color context
Green bands appear lighter when bounded by orange, and darker next to red. In fact they're exactly the same shade. The illusion illustrates brightness assimilation, which causes color perception to vary according to the background. Conversely, color constancy enables us to perceive a shirt as white whether it's lit by sunlight or by incandescent or fluorescent lamps.

In perspective ▶
Artists take advantage of the way the brain processes visual information in order to create the illusion of depth. The image at right is obviously two-dimensional. But the tub and shower "appear" farther away because the tiles get smaller.

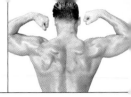

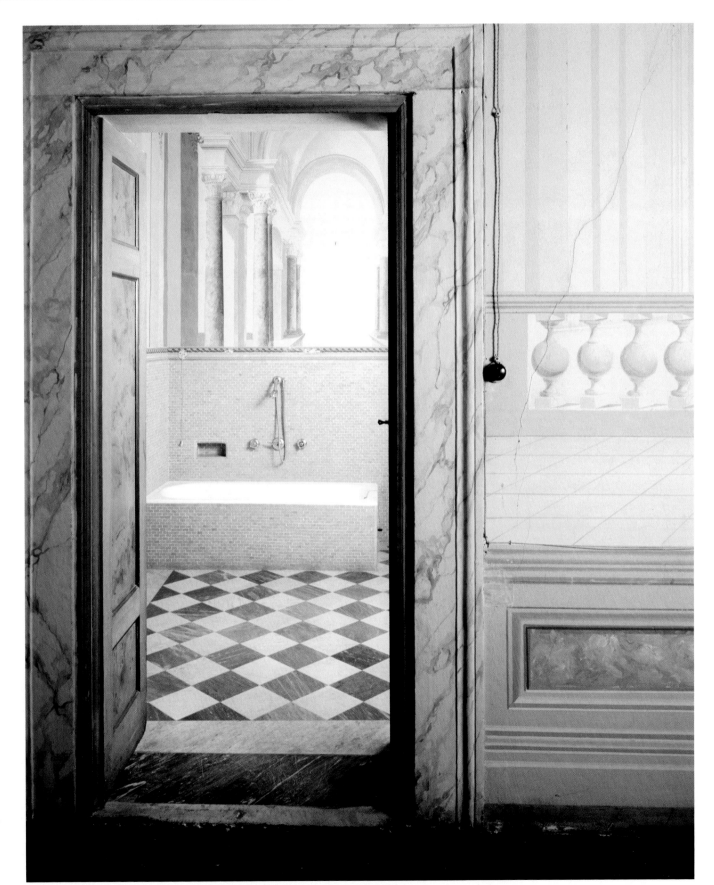

SEE ALSO

Light of
many colors
102-103

Basics of
the brain
244-245

Seeing and
believing
246-247

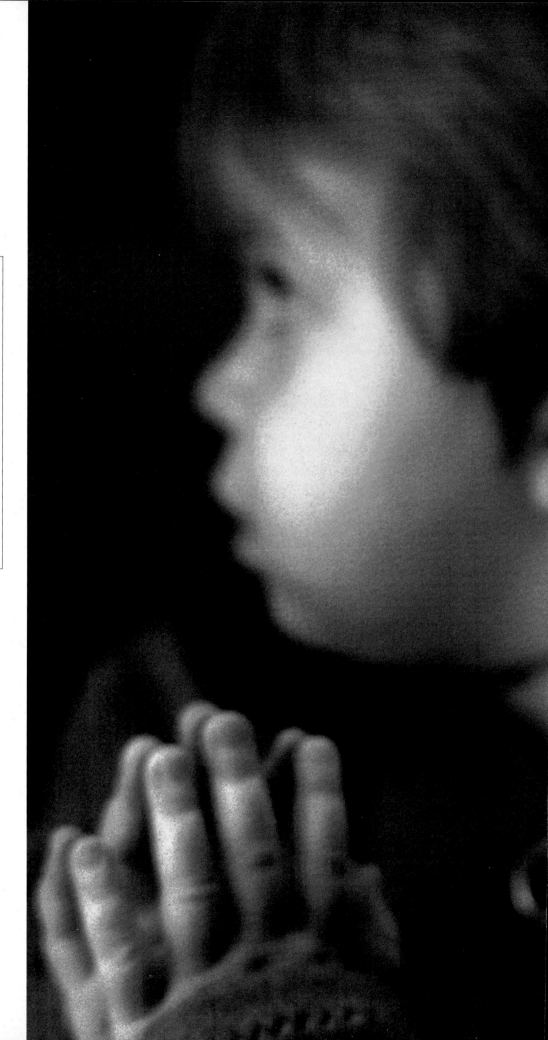

Who's there?

Very little in the human brain is "hard-wired" at birth. Nearly every circuit involving vision, perception, and interpretation of the environment assembles itself through repeated sensory inputs and the familiar process of trial and error. That's one reason that it takes an infant about a year and a half to recognize the image in the mirror as him- or herself. Making that connection is a non-trivial task that may be related to cognitive capacity. Our closest genetic relatives, the great apes, also recognize themselves. A recent study suggests that the bottlenose dolphin has become the first non-primate to pass the mirror test of self-recognition.

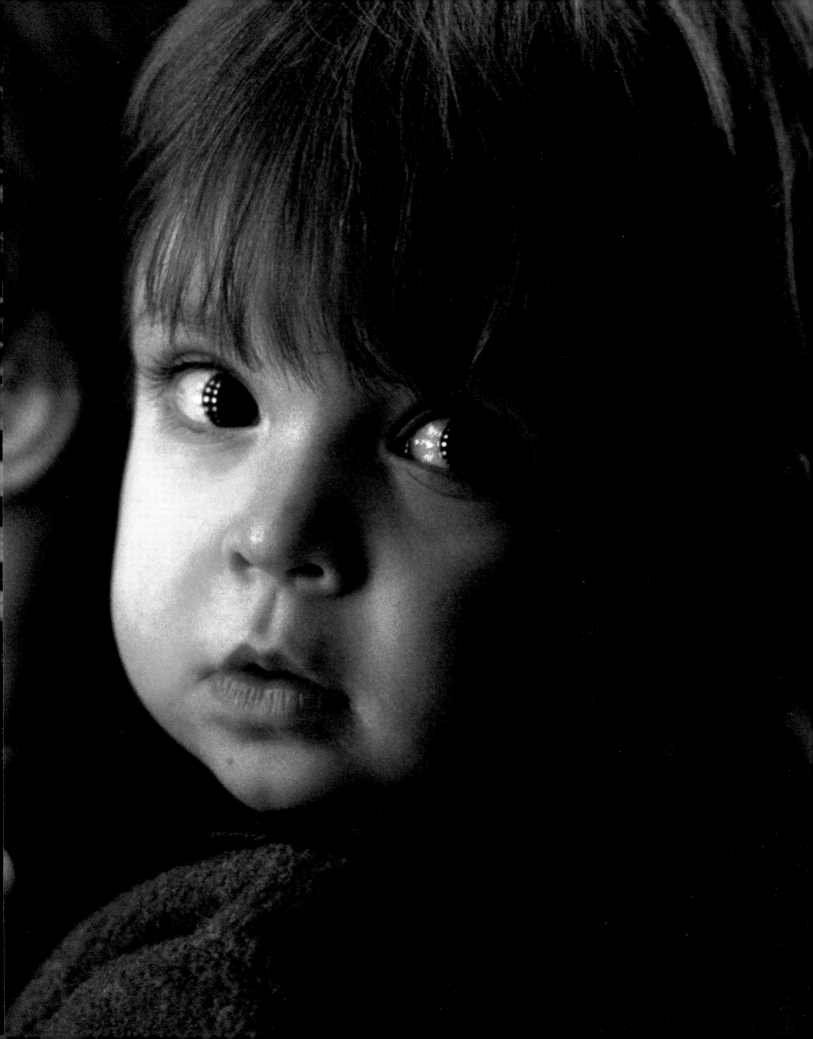

Sounds of a lifetime

The ear and eye are kindred organs. Not only do they sense stimuli, they also help gauge both distance and direction. That's why they come in pairs.

The human ear is no less astonishing than the eye, both in its range and in its subtlety of perception. It can discriminate frequencies between 20 and 20,000 cycles per second. It can detect the rustling of leaves but also handle the overpowering vibrational roar of a nearby jet taking off. Like the eye, the ear is a transducer: Something that turns one form of energy into another.

The ear turns mechanical force from alternating waves of compression and rarefaction in the air into electrical signals. As sound waves strike the eardrum, they cause it to move back and forth. That motion is carried by a set of three linked, leverlike bones in the middle ear to a membrane at the end of a fluid-filled spiral enclosure called the cochlea.

This membrane is much smaller than the eardrum, and the bony levers multiply incoming forces. So the pressure exerted on the cochlea is about 20 times stronger than that striking the eardrum. It alternately compresses and relaxes the inner-ear fluid at exactly the same frequency as the incoming sound, causing tiny hairs in the cochlea to move back and forth in time with the pressure. These hairs are mechanically gated; that is, when they bend, ion channels open, allowing charges to pass through that depolarize the cells. Such electrical signals are transferred to auditory nerve fibers that carry them to the brain's temporal lobes for translation and analysis. ∎

Too loud ▶
Because hearing is such an intensely mechanical process, it wears out a bit as you age. Exposure to sounds over 120 decibels can hasten the process. As long as some cochlear hairs and neurons remain healthy, a hearing aid may help. It amplifies sound and transfers it to a membrane or directly to the cochlea through vibrations in nearby bone.

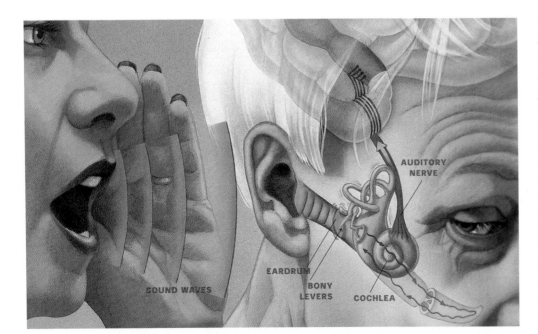

◀ Not only volume
Knowing a sound's direction is important. That process occurs in the brain, not the ear. Structures in the auditory cortex perceive when the same sound arrives slightly earlier and a bit stronger at one ear than at the other. Your brain uses that difference to calculate an approximate location.

AUDITORY NERVE

SOUND WAVES

EARDRUM

BONY LEVERS

COCHLEA

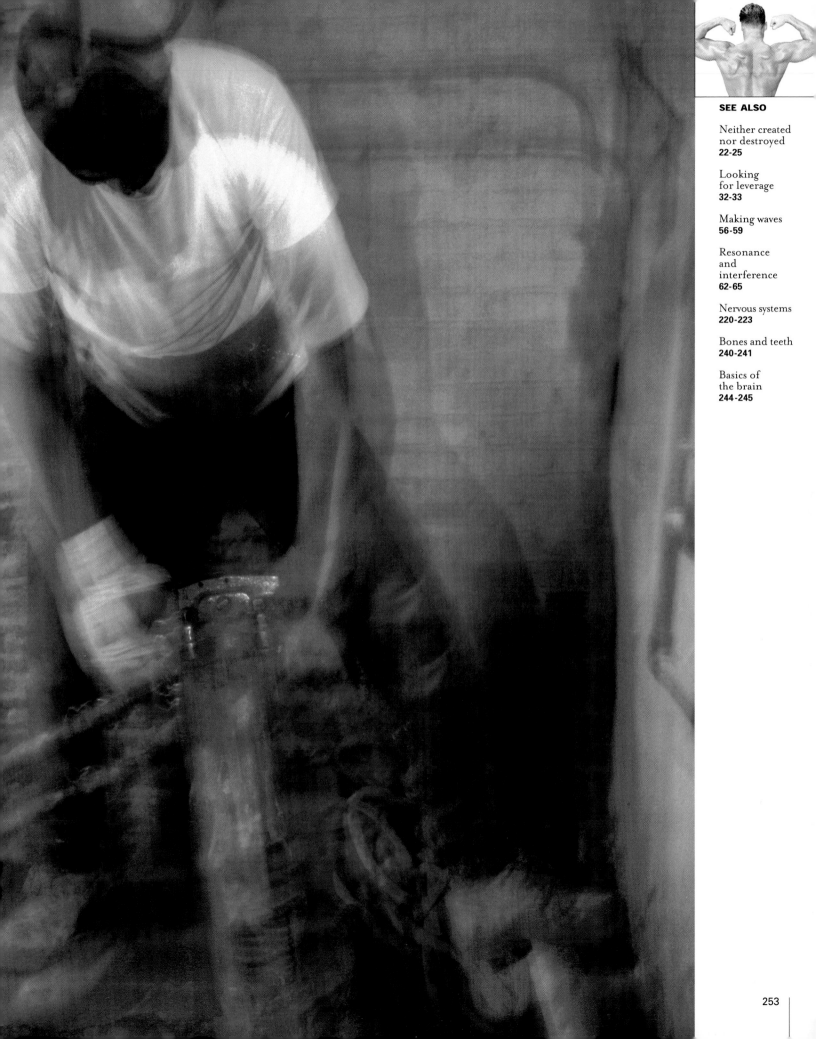

Taste and smell

Flavors of the world, unite! Our senses of taste and smell are so tied to each other we often confuse them—until we get a severe cold or sinus condition.

When our primate ancestors first took to the trees, thereby elevating their snouts off the ground, our sense of smell became much less important. Compared to the olfactory equipment of an ordinary dog or cat, your nose is pathetic. But it's still capable of detecting chemicals in the parts-per-million range.

Each nostril contains as many as ten million olfactory receptor cells, located just below the bony plate that separates your nasal cavity from your brain, and far enough out of the way that you have to sniff a little to really get them going. Each has a number of hairlike cilia that dangle down through a mucus membrane and snare passing molecules. They detect only the water-soluble ones because, like your tongue, your nose only senses dissolved substances.

If the molecules fit into the specialized attachment sites on one of the cilia, sodium and potassium ion channels open and the nerve cell fires, sending its signal across the bone plate and into a center called the olfactory bulb. Your nose works hard, and it gets special treatment: Its sensory receptors are attached to the only nerve cells in your body that regularly divide to replace themselves, about every two months. ■

Primacy of smell

Estimated to be about 10,000 times more sensitive than taste, the human sense of smell varies widely. Most people distinguish thousands of different odors. Certain smells can trigger surprisingly vivid memories or intense feelings—possibly because the olfactory nerves are hard-wired directly into the limbic system, the center of emotion (below). This profound relationship between scents and sensibility helps fuel the multibillion-dollar fragrance and deodorant industries.

When taste comes to smell ▶
You've got some 10,000 taste buds, each equipped with about 50 receptor cells. Those receptors, accessible to fluids through circular taste pores, can distinguish only four basic tastes: sweet, sour, bitter, and salty. Different proportions produce unique flavors.

LIMBIC SYSTEM

OLFACTORY NERVE

RECEPTOR CELLS

Superior snouts ▶
Think you have a good nose? It's nothing compared to the average pooch. It is estimated that dogs have around 200 million olfactory receptors, making their sense of smell thousands of times keener than ours. Of course, we're not exactly untalented. Our noses can detect some odors (such as the rotten-garlic smell added for safety reasons to odorless natural gas) at concentrations of one part in 50 million. But that's not good enough to track a fugitive or sniff out drugs.

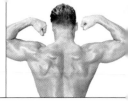

SEE ALSO

The death
of a cell
198-199

Nervous systems
220-223

Basics of
the brain
244-245

Emotion
and behavior
260-261

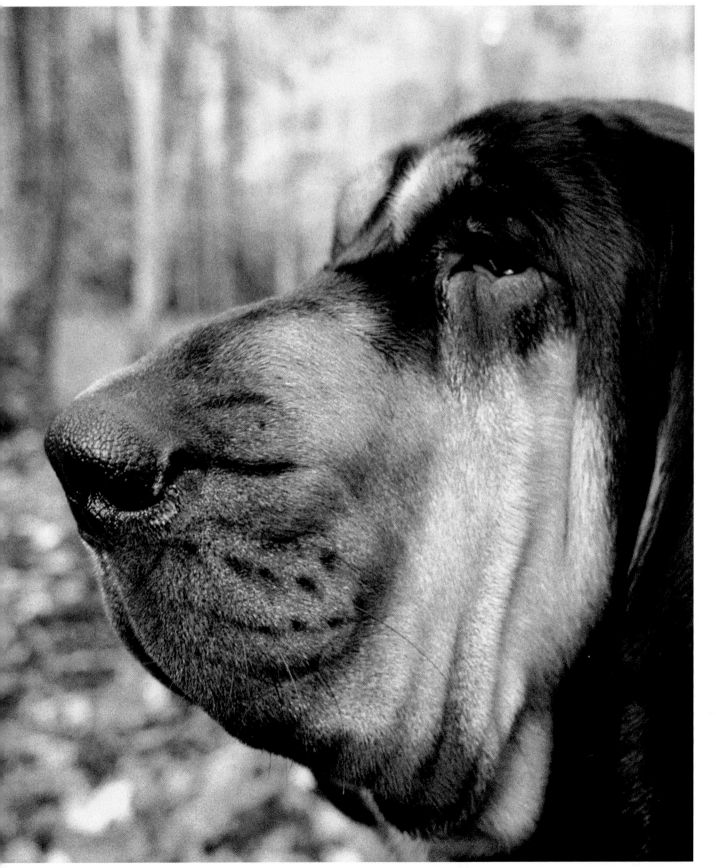

Thresholds of pain

You might say pain and pleasure keep us in touch with the world. But both sensations continue to surprise, no matter how common.

Pain is nature's way of getting your attention, and it pretty much works every time. It is conveyed by a completely different pathway from the one that relays the touch signals of a soothing massage. Pain is caused by the response of several million naked nerve endings arrayed throughout your skin and in most parts of the body. Oddly, there are no pain receptors in the brain, despite its billions of nerve cells.

Pain receptors are of three types. One reacts to mechanical damage, another to temperature changes. The third—called polymodal—responds to a variety of stimuli such as the chemical insults that occur when fluids build up during inflammation. Mechanical and thermal signals shoot to the 31 pairs of major nerves in the spinal cord on a neural fast track made of fibers that transmit at the rate of 30 meters per second. Polymodal signals run on slower fibers at a relatively trifling 12 meters per second. Both were activated the last time you cut yourself: You may recall first a sharp twinge, then a duller but more persistent pain.

The brain has natural painkillers in the form of endorphins, so named because they are considered "**endo**genous mo**rphin**es." They can arise during or after exercise (you may have experienced them as "runner's high") or other short-term stresses. ■

Pain delayed: mind over matter? ▶
Distress and anxiety can often heighten the acute pain of a sudden injury, such as this soccer player's mishap. Conversely, the brain often responds to serious wounds by producing more endorphins: natural chemicals that, like morphine, suppress pain.

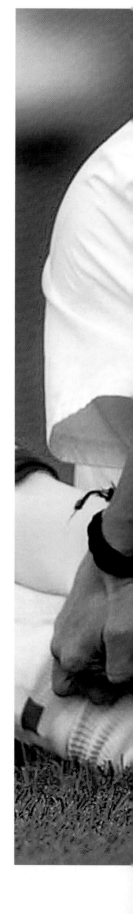

Over-the-counter painkillers

Much of the pain of modern life has been banished by popular analgesics. Aspirin, the oldest, works by inhibiting production of a particular type of potent, hormone-like fatty acids called prostaglandins. Some aid reproduction, help regulate body temperature, or assist in other chores. But one kind makes pain more painful: It multiplies the receptor response, enhances inflammation, and brings on fever by instructing the hypothalamus to raise its "set point," much the way we turn up a thermostat. By blocking the manufacture of these prostaglandins, aspirin relieves inflammation. So does ibuprofen—and its close relatives known as naproxen and fenoprofen.

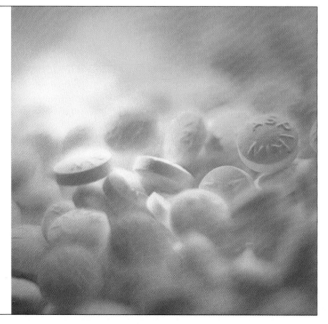

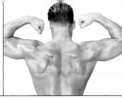

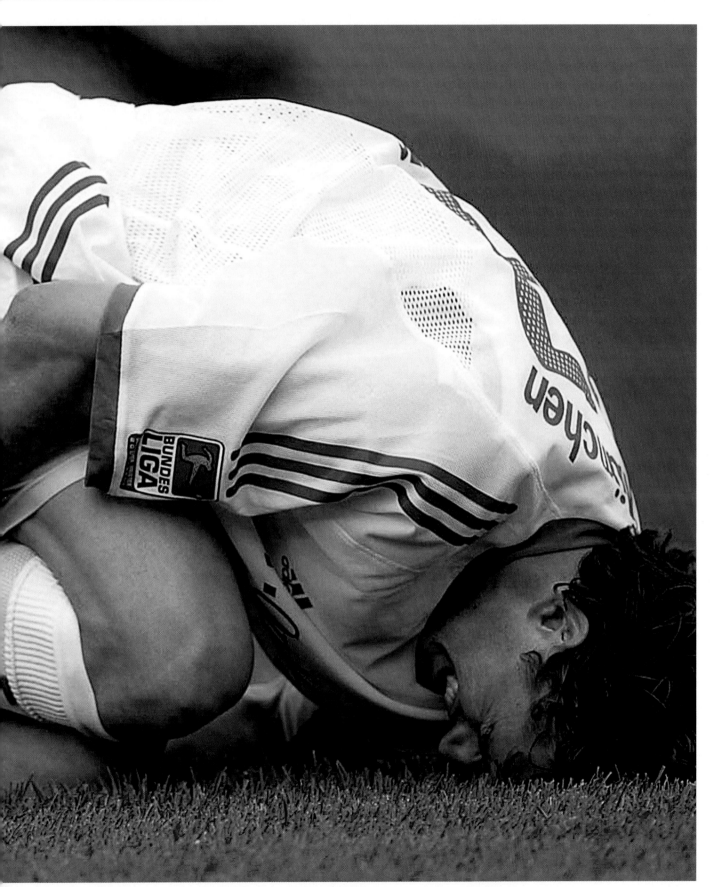

Suppressing pain

Some who suffer chronic pain can be trained to ignore part or all of it, and the science of pain management has become a complex and growing field, combining drug regimens, invasive procedures, and psychological techniques. Some treatments are drawn from "alternative medicine" sources such as acupuncture; others harness the kind of mental discipline shown by this firewalker in northern India. Many of these performers claim they feel no pain during their inflammatory excursions. This one doused his garments with water just before plunging into the flames.

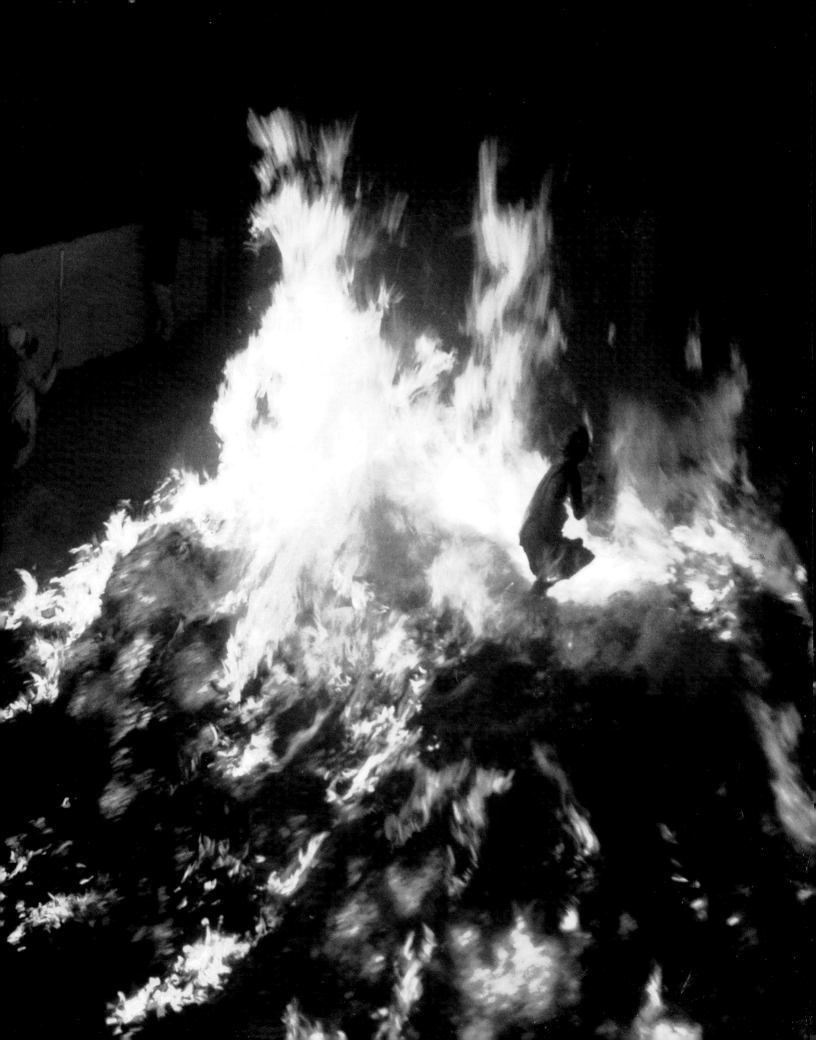

Emotion and behavior

Modern science is revealing the biochemical basis of both normal feelings and the extreme or destructive emotions of mental disorders.

Emotions inevitably come down to chemistry. Even the profound grief of these French citizens witnessing the Nazi invasion of Paris in 1941 (right) can be explained on a purely chemical basis: Their deep emotional states are part of a normal reaction to threats, grief, and horror, involving different levels of various neurotransmitters at different receptors. As knowledge of these chemical messengers has increased, scientists have been able to influence the brain's activity and signals—and thus to treat many forms of mental illness, both acute and chronic, that once were considered hopelessly irremediable.

Depression, humanity's most common mental-health problem, seems to result from inadequate supplies of at least two neurotransmitters: serotonin and norepinephrine. Antidepressant drugs are designed to increase their abundance in the brain. Severe anxiety disorders may be treated with a family of tranquilizers called benzodiazepines, such as Valium and Xanax. These drugs stimulate receptor sites for a particular neurotransmitter called GABA , which makes neurons less excitable. When that occurs in the limbic system, an immediate reduction in anxiety and fear results. ■

Why cry? ▶
Homo sapiens *is the only species that weeps. And despite decades of study, it is still not clear what the body is trying to achieve with the production of emotional tears. The salty liquid contains numerous substances that have a helpful effect on eye chemistry, as well as natural defenses against bacteria. But why sadness produces excessive tears remains unknown.*

Getting attention

Amphetamines and other stimulants, once prescribed for various maladies, were largely discontinued once their potential for addiction and abuse became clear. Now their most common use is in controlling attention-deficit hyperactivity disorder (ADHD), a condition that affects 3 to 5 percent of American schoolchildren—and can continue into adulthood. ADHD is substantially more prevalent in boys. Thousands rely on Ritalin, a drug that boosts production of two key neurotransmitters, norepinephrine and dopamine. Acting together, these two substances enhance concentration and alertness while stabilizing both mood and speech activity. Widespread use of such drugs is controversial. But in properly diagnosed children, there is considerable evidence that effects can reduce aggressive behavior and increase learning ability dramatically.

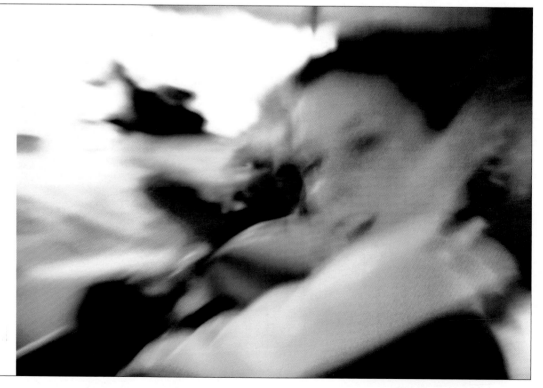

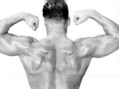

SEE ALSO

Nervous systems
220-223

Basics of
the brain
244-245

Taste and smell
254-255

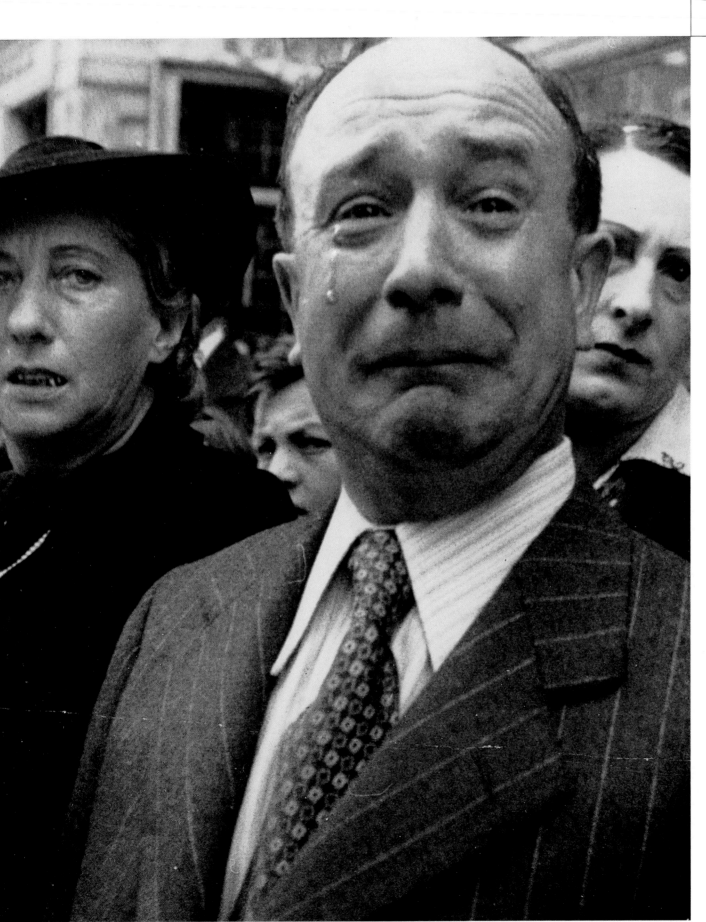

Drugs of pleasure

Nearly all of us take drugs that affect our minds and moods. It's just that most of the time, they're legal.

f you're like many Americans, you're a habitual drug user—loading up on caffeine, nicotine, or ethyl alcohol. The effects they produce, at least when taken in moderation, are socially acceptable. Caffeine, the world's most popular drug, is one of three related stimulants derived from plants. The others are theobromine, found in cacao, and theophylline, present in tea (as is caffeine).

Even at doses as low as 100 milligrams—the average amount in a single cup of coffee—caffeine improves intellectual and physical performance during the three to five hours that it is most active. It also boosts stomach acid production, increases heart output, and dilates the coronary arteries. In the brain, however, it constricts blood vessels. Heavy coffee drinkers who go "cold turkey" may experience unaccustomed dilation of the cerebral blood vessels in the form of severe headaches; conversely, caffeine can provide relief for those with migraines.

Ethanol—grain alcohol—acts on at least four of the major neurotransmitters. It enhances action of your brain's biggest downer, the inhibitory substance called GABA, while damping the function of glutamate, the major excitatory neutrotransmitter. It also appears to stimulate the effect of dopamine pleasure centers and decrease the activity of impulse-restraining serotonin, often prompting inappropriate aggression. Cheers. ■

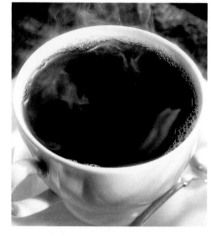

Hot stuff ▲
Caffeine works in the brain by blocking receptor sites ordinarily occupied by adenosine. Adenosine has a sedative effect, and normally inhibits the release of a host of neurotransmitters that includes norepinephrine, dopamine, and acetylcholine. When adenosine can't act, those substances go into wider circulation, prompting body-wide stimulation.

Cigarette suckers ▶
Nicotine, the highly addictive substance found in tobacco, is so toxic that only 60 milligrams—an amount roughly equal to one-fifth of an aspirin tablet—would kill you on the spot. That's why kids trying their first cigarettes can become nauseated. Tolerance develops over time. Most cigarettes contain from 0.5 to 2.0 milligrams of nicotine; about 20 percent reaches the bloodstream. Nicotine acts on acetylcholine receptors similar to those that prompt muscles to contract. It can temporarily aid concentration and motor nerve performance, and it increases heart rate while damping nerve signals from muscles, perhaps causing a sense of relaxation.

Love and chocolate

The most romantic of sweets has long had a reputation for modifying the mind. Explorers in the New World raved about the stimulant effects of drinks made from cacao "beans." Now researchers know that chocolate contains caffeinelike theobromine and hundreds of other compounds, including phenylethylamine, which is produced naturally by the nervous system and can elevate one's mood much as amphetamines do. Indeed, phenylethylamine may even be a key ingredient in the chemistry of falling in love. It's also known that depressed people often binge on chocolate.

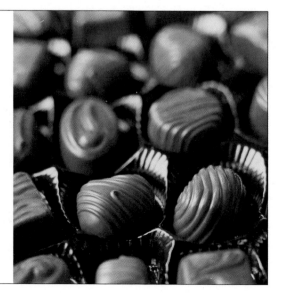

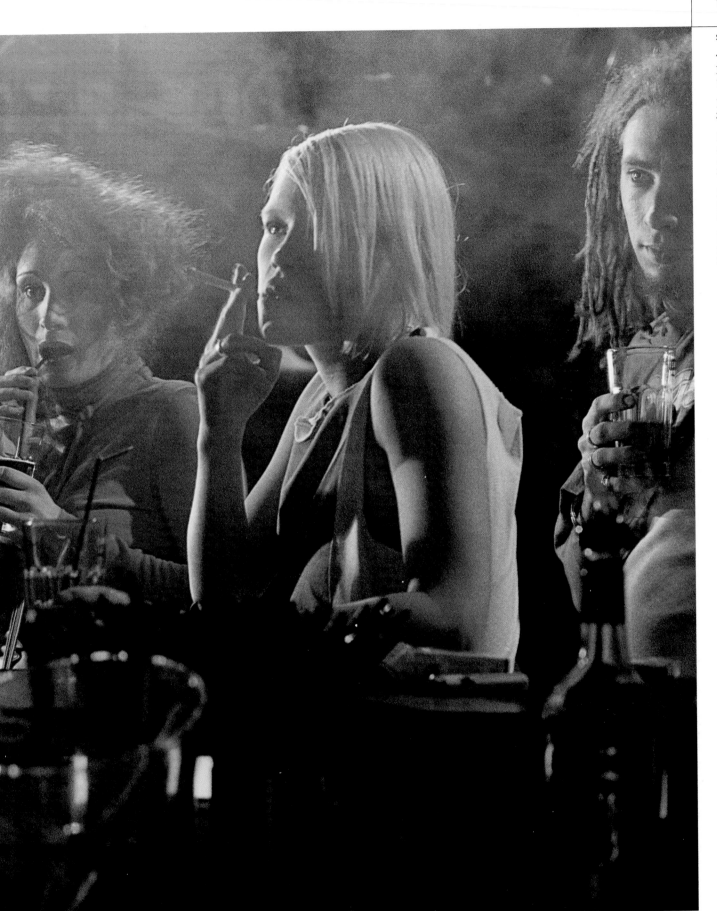

Sleep and dreams

Your brain never rests. Not when you sit or lie quietly, not when you meditate, not even when you sleep. And certainly not while you dream.

We spend about a third of our lives in the realm of sleep. Why do we do it—and what do we do while we're there? The seemingly obvious answer is that the body is repairing itself. But that just isn't so. In fact, your tissues can get all the maintenance they need while you're flopped on the sofa watching television. Believe it or not, you can function fairly well after 48 or even 60 hours without sleep—physically, anyway. Your mental concentration, motor ability, and social skills may decline; but you don't fall to pieces.

So what's it for? Sleep seems to be the brain's way of putting its own house in order. Hormone levels are recharged during sleep. Large, predatory animals with relatively more cortex (such as humans and cats) sleep more than small prey with less cortex (mice or fish). More sophisticated brains also tend to engage in rapid eye movement (REM) sleep—in which dreams occur—as well as deeper, non-REM sleep. On an average night, you'll pass through several stages on your way to dreamland, as your body temperature drops about two degrees. You'll also probably change body position as you switch sleep states. During non-REM sleep, most folks shift six or so times an hour.

Sleep, therefore, is hardly a passive activity. Indeed, studies have shown that your brain may have an even higher oxygen-uptake rate during some sleep stages than it does when you're fully awake. And when monitored by electroencephalograms (EEGs), the brain generates the same kind of waves during REM sleep as it does when you are fully awake and alert. That's a lot of action. It's amazing that we don't wake up exhausted. ■

Now I lay me down ▶
EEGs show that there are two basic kinds of sleep: "slow-wave" periods followed by the brain-wave speed-up of REM sleep. Usually people cycle through four progressively deeper stages of slow-wave sleep over the course of a half-hour or so, before hitting the REM phase. REM episodes typically last about 10 or 15 minutes. Then the cycle repeats itself.

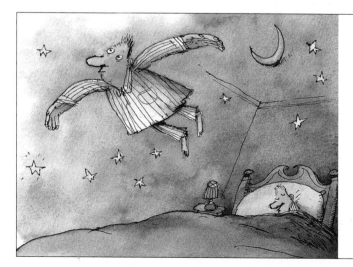

Why dream?

If allowed to engage in non-REM sleep but awakened before entering REM, people can get anxious and irritable—although they're rested in terms of the time they've spent unconscious. That suggests that dreams may have a purpose. But what? Some think it's the brain's way of dumping useless sensory data. Others believe the brain is organizing data into patterns for comprehension. That may be why newborns spend half of their sleep time in REM, while adults devote as little as 20 percent.

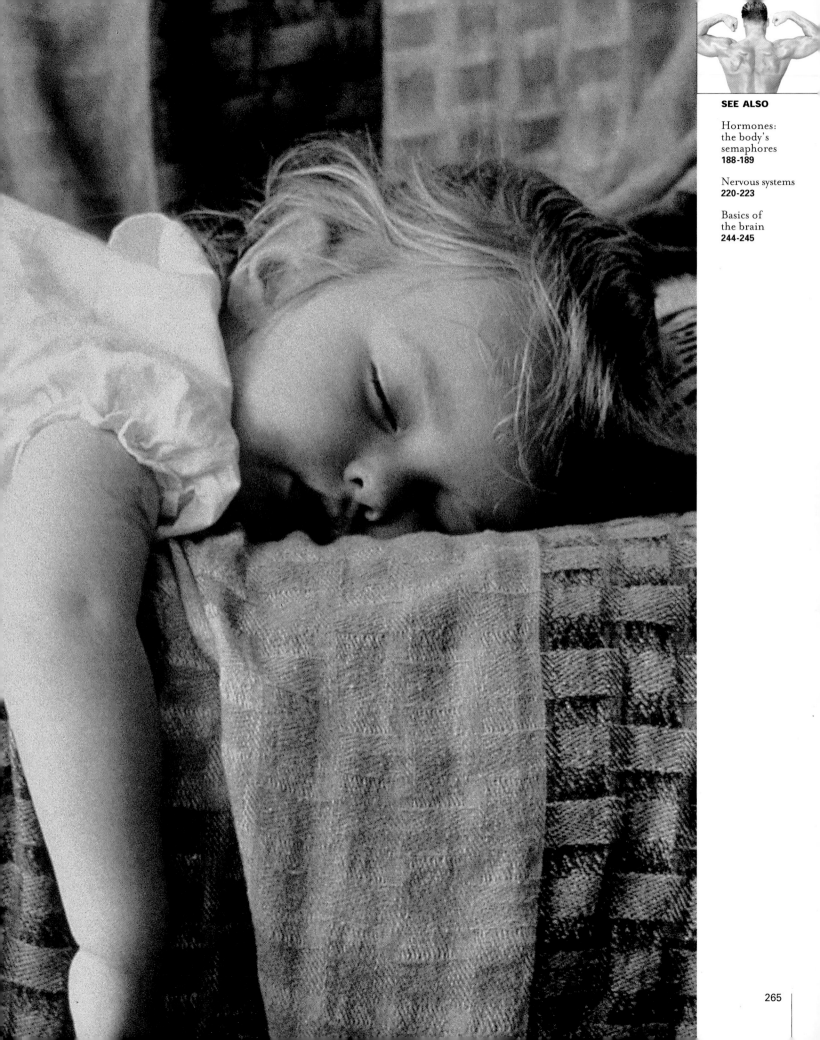

SEE ALSO

Hormones:
the body's
semaphores
188-189

Nervous systems
220-223

Basics of
the brain
244-245

Intimations of mortality

While life is a downhill slide, the angle often varies greatly with lifestyle and genes. And as each year passes by, life expectancy doesn't drop, it increases.

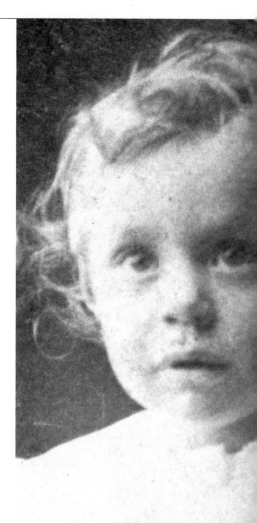

You may justifiably feel that you're in the prime of life. But physiologically, you peaked somewhere around the age of 20, and you're already embarked on the Long Goodbye. You can see signs in the mirror, perhaps, or in a collage of photos taken over time (right). With aging, skin gets thinner, drier, less elastic—and more wrinkled. The nose usually lengthens and widens significantly; ears get larger. Hair thins and turns gray, as follicle cells quit providing pigment, or it falls out altogether. By the age of 70, body height shrinks by perhaps two inches, as gravity compresses spinal disks and joints. Even the taste buds wither, decreasing in number by two-thirds.

Just why all this happens is not clear. One theory holds that many cells simply can't reproduce themselves indefinitely. Cell culture studies have shown that, for humans over 70 years old, somewhere between 40 and 60 cell division cycles may be the upper limit. You can fight the natural flow of things with good diet and plenty of exercise. Those who stay in shape can have, at age 60, as much as 80 percent of the strength that they had at 25. But the best defense may simply be to remember that some things can't be changed: In aging, as in all biological processes, nature bats last. ■

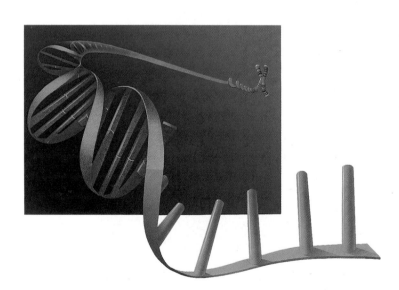

◄ Telomeres

At the end of each chromosome is a telomere, which does not code for genes yet is essential for chromosomal survival. Whenever a cell replicates, enzymes always fail to translate the last bases on the telomere. When a telomere gets too short, the cell stops dividing and dies. It's all part of the aging process.

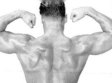

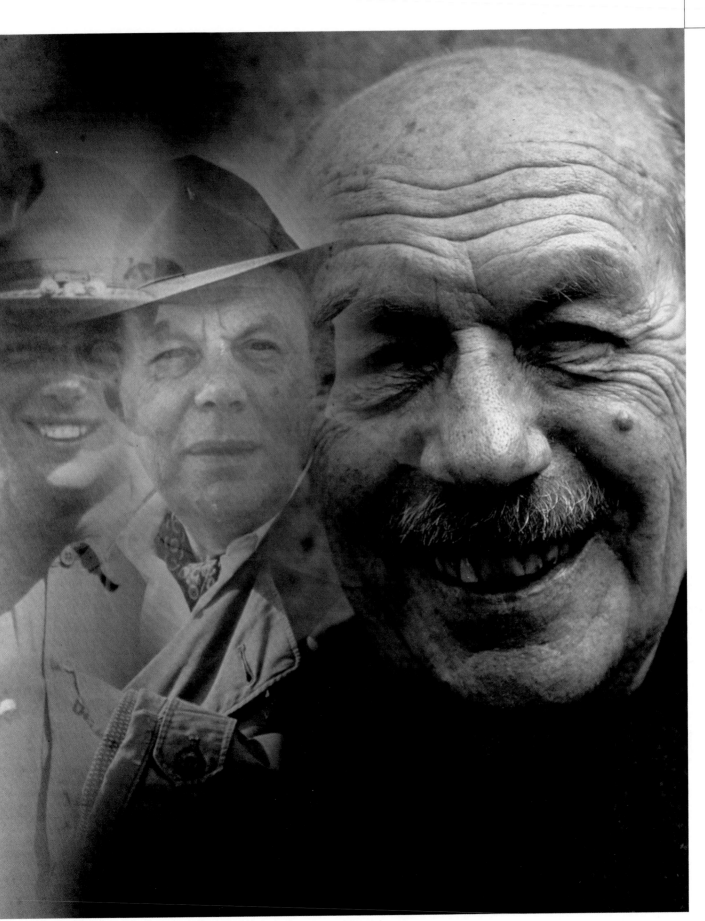

SEE ALSO

Getting started
14-15

Getting going
182-183

Evolution:
nature's
test drives
196-197

The death
of a cell
198-199

Additional Reading

David Bodanis, *The Body Book*; Douglas Giancoli, *Physics*, 4th ed.; David Macaulay, *The Way Things Work*; Macmillan, *The Way Science Works*; Macmillan, *The Way Nature Works*; Elaine N. Marieb, *Essentials of Human Anatomy & Physiology*, 4th ed.; Nina Morgan, *Chemistry in Action: the Molecules of Everyday Life*; National Geographic Society, *The Incredible Machine*; *Oxford Illustrated Encyclopedia of Invention and Technology*; Reader's Digest, *How in the World?*; Reader's Digest, *ABC's of the Human Body*; Richard Restak, *The Brain*; Colin A. Ronan, ed., *Science Explained: the World of Science in Everyday Life*; Lauralee Sherwood, *Human Physiology: From Cells to Systems*, 2nd ed.; *Scientific American Library Series*; Time-Life Books, *Understanding Science and Nature Series*; Times Books, *The Way Life Works*; James Trefil and Robert M. Hazen, *The Sciences, An Integrated Approach*; James Trefil and Robert M. Hazen, *Science Matters: Achieving Scientific Literacy*; Philip Whitfield, *The Human Body Explained*

Acknowledgments

The Book Division is grateful to many individuals who contributed to *The New Everyday Science Explained*. Our special thanks goes to: Charlotte M. Agnone, M.D., Andrew R. Baden, Ph.D., Elizabeth H. Blackburn, Ph.D., Peter S. Birk, M.D., Edson C. Brolin, M.S, Greer Garrett, M.A., Carol E. Horn, M.D., Vinod K. Jain, Ph.D., Alexandra Littlehales, Paul D. Lowman, Jr., Ph.D., PhotoAssist, Roger A. Pielke, Sr., Ph.D., Lorraine C. Racusen, Ph.D., Richard Racusen, Ph.D., Richard M. Restak, M.D., Masoud Sanayei, Ph.D., Brian E. Strass, Haider Taha, Ph.D., James Trefil.

One of the world's largest nonprofit scientific and educational organizations, the National Geographic Society was founded in 1888 "for the increase and diffusion of geographic knowledge." Fulfilling this mission, the Society educates and inspires millions every day through its magazines, books, television programs, videos, maps and atlases, research grants, the National Geographic Bee, teacher workshops, and innovative classroom materials. The Society is supported through membership dues, charitable gifts, and income from the sale of its educational products. This support is vital to National Geographic's mission to increase global understanding and promote conservation of our planet through exploration, research, and education.

For more information, please call
1-800-NGS LINE (647-5463)
or write to the following address:

National Geographic Society
1145 17th Street N.W.
Washington, D.C. 20036-4688
U.S.A.

Visit the Society's Web site at www.nationalgeographic.com.

Published by the National Geographic Society
John M. Fahey, Jr. *President and Chief Executive Officer*
Gilbert M. Grosvenor *Chairman of the Board*
Nina D. Hoffman *Executive Vice President*

Prepared by the Book Division
Kevin Mulroy *Vice President and Editor-in-Chief*
Charles Kogod *Illustrations Director*
Marianne R. Koszorus *Design Director*

Staff for *The New Everyday Science Explained*
Barbara Brownell-Grogan *Executive Editor*
Susan Blair *Project and Illustrations Editor*
Cinda Rose *Art Director*
Rebecca Lescaze *Text Editor*
Sharon K. Berry *Illustrations Specialist*
Gary Colbert *Production Director*
Lewis Bassford *Production Project Manager*
Susan Hitchcock *Contributing Editor*
Connie D. Binder *Indexer*

Manufacturing and Quality Control
Christopher A. Liedel *Chief Financial Officer*
Phillip L. Schlosser *Managing Director*
John T. Dunn *Technical Director*
Maryclare McGinty *Manager*

Composition for this book by the National Geographic Book Division. Printed and bound by R. R. Donnelly & Sons, Willard, Ohio. Color separations by Quad Imaging, Alexandria, Virginia. Dust jacket printed by the Miken Co., Cheektowaga, New York.

Library of Congress Cataloging-in-Publication Data

Suplee, Curt.
The new evceryday science explained / by Curt Suplee.
p. cm.
ISBN 0-7922-7357-5 (alk.paper)
I. Science—Popular works. 2. Physics—Popular works. 3. Chemistry—Popular works. 4. Human biology—Popular works. I. Title

Q162.S943 2004
500—dc22

2003061445

Credits

Cover Digital Vision/gettyimages

Frontmatter

1 Lester Lefkowitz/CORBIS; 2-3 Matt Meadows/Peter Arnold Inc.; 4 (le) Comstock IMAGES; 4 (rt) Photodisc Green/gettyimages; 5 (both) CORBIS; 6-7 Leslie Sponseller/Stone/002gettyimages

Matter and Motion

8-9 Insurance Institute for Highway Safety; 10 Richard Hutchings/CORBIS; 11 Randy Lincks/CORBIS; 12-13 Sally J. Bensusen Visual Science Studio; 14 Bruce Dale; 14-15 Tony Duffy/Allsport/gettyimages; 15 (lo) Richard Thompson; 16 The Image Bank/gettyimages; 17 David Madison Sports Images; 18 Peter Lloyd; 19 (up) David Madison Sports Images; 19 (lo) Paul Barton/The Stock Market/CORBIS; 20 Richard Thompson; 21 NASA; 22 Runk/Schoenberger/Grant Heilman Photography; 23 Adam Woolfitt/CORBIS; 24-5 Robert Landau/West Light/CORBIS; 26-7 Vandystadt/Allsport/gettyimages; 27 (le) Richard Thompson; 27 (rt) David Madison Sports Images; 28 Leonard Lessin/Peter Arnold Inc.; 29 David Madison Sport Images; 30 (up) Barry L. Runk/Grant Heilman Photography; 30 (lo) Richard Thompson; 31 Duomo Photography Inc; 32 Richard Thompson; 32-3 Wood Ronsaville Harlin, Inc.; 33 Sisse Brimberg; 34 James P. Blair; 35 (up) Stone/gettyimages; 35 (lo) James L. Amos; 36-7 Lawrence Manning/CORBIS; 38 CORBIS; 38-9 The Image Bank/gettyimages; 40 (up) Michael & Patricia Fogden/CORBIS; 40 (lo) SPL/Photo Researchers; 41 Yoav Levy/Phototake; 42 Science Photo Library/Photo Researchers; 42-3 Dr. Jeremy Burgess/Photo Researchers; 44 (both) Runk/Schoenberger/Grant Heilman Photography; 45 Rick Gayle/CORBIS; 46-7 Outdoor Images; 48 E.R. Degginger /Animals Animals; 48-9 Pete Saloutos/The Stock Market/CORBIS; 50 Robert W. Tope, Natural Science Illustrations; 50-1 AP/Wide World Photos; 52 The Image Bank/gettyimages; 52-3 Donald Miralle/gettyimages; 54 Sally J. Bensusen Visual Science Studio; 54-5 Mark Thompson/gettyimages; 55 Sally J. Bensusen Visual Science Studio; 56 (up) (photo) Stone/gettyimages; (art) Theophilus Britt Griswold; 56-7 (lo) Sally J. Bensusen Visual Science Studio; 57 (up) Chase Swift/CORBIS; 58-9 Maxell; 60 Robert W. Tope, Natural Science Illustrations; 61 CORBIS; 62 Wood Ronsaville Harlin, Inc.; 63 Jeri Gleiter/ Peter Arnold Inc.; 64 MSCUA University of Washington Libraries Neg. #4; 64-5 MSCUA University of Washington Libraries Neg. #12; 66 Wood Ronsaville Harlin, Inc.; 66-7 NASA; 68 Richard Thompson; 69 James Noble/CORBIS; 70 Sally J. Bensusen Visual Science Studio; 70-1 Bill Ross/CORBIS; 72 (up) NASA/Photo Researchers; 72 (lo) Imagery Index Stock; 73 NASA

Forces of Nature

74-5 Roger Ressmeyer; 76 Richard Thompson; 77 Runk/Schoenberger/Grant Heilman Photography; 78 Pierre Mion; 78-9 Stuart Franklin/gettyimages; 80 Richard Thompson; 80-1 R. Berenholtz/ The Stock Market/CORBIS; 81 Denis Boulanger/Allsport/gettyimages; 82 (up) Robert W. Tope, Natural Science Illustrations; 82 (lo) AFP/CORBIS; 83 Douglas Peebles/CORBIS; 84-5 Sally J. Bensusen Visual Science Studio; 86 Craig Kiefer; 86-7 Stone/gettyimages; 87 Richard Thompson; 88-9 Chris Johns NGP; 90-1 Wood Ronsaville Harlin, Inc.; 92 Charles Falco/Photo Researchers; 93 James A. Sugar; 94 R. Curtis/Photo Researchers; 94-5 Richard Thompson; 96-7 NASA; 98-9 Lester Lefkowitz/CORBIS; 100-1 (up) Liaison Agency/gettyimages; 100-1 (lo) Theophilus Britt Griswold; 102 Claude Nuridsany & Marie Perennou/Photo Researchers; 103 (up) The Image Bank/gettyimages; 103 (lo) Taxi/gettyimages; 104 (up) Thinkstock/gettyimages; 104 (lo) James Stuart: Beaford Archive/CORBIS; 104-5 The Image Bank/gettyimages; 106 Terranova Int'l/Photo Researchers; 106-7 Sally J. Bensusen Visual Science Studio; 108 (up) Frans Lanting/ Minden Pictures; 108 (lo) Wood Ronsaville Harlin, Inc.; 109 Hans Pfletschinger/Peter Arnold Inc.; 110 Lester Lefkowitz/CORBIS; 111 (up) SPL/Photo Researchers; 111 (lo) Dr. R. J. Allen et al/Photo Researchers; 112-3 The Image Bank/gettyimages; 114 Richard Thompson; 114-5 Dean Conger/CORBIS; 115 Phototake; 116 Wood Ronsaville Harlin, Inc.; 116-7 John Madere/CORBIS; 118-9 Ted Horowitz/CORBIS; 118 Photodisc Green/gettyimages; 120-1 Dr. Jeremy Burgess/Photo Researchers; 122 (up rt) Photo Researchers; 122 (lo le) Stone/gettyimages; 123 Taxi/gettyimages; 124 IMP Team JPL NASA; 125 Bembaron Jeremy/CORBIS SYGMA; 126 Scott Thorn Barrows/Craig Keifer; 126-7 Jim Zuckerman/CORBIS

The Right Stuff

128-9 UPI/Bettmann/CORBIS; 130 (up) Photodisc Green/gettyimages; 130 (lo) Richard Thompson; 131 Photodisc Green/gettyimages; 132 (up) Barry L. Runk/Grant Heilman Photography; 132 (lo) Craig Kiefer; 133 Photographer's Choice/gettyimages; 134 (le) Richard Thompson; 134 (rt) Charles D. Winters/Photo Researchers; 135 E. Schrempp/Photo Researchers; 136 H. Yang (UIUC)/J. Hester (ASU)/NASA; 136-7 Sally J. Bensusen Visual Science Studio; 138 (up) Craig Kiefer; 138 (lo) James L. Amos; 138 (ctr) Craig Kiefer; 139 Jose Fuste Raga/CORBIS; 140-1 James L. Amos; 142 H. David Seawall/CORBIS; 142-3 George Steinmetz; 144 Richard Thompson; 145 (inset) Craig Kiefer; 145 Paul Chesley; 146-7 Robert W. Tope, Natural Science Illustrations; 147 The Image Bank/gettyimages; 148-9 Don King Films; 150 Jim Winkley/Ecoscene/CORBIS; 150-1 Theophilus Britt Griswold; 151 Ted Spiegel/CORBIS; 152 Adam Wooflitt/CORBIS; 152-3 Lloyd Cluff/CORBIS; 154 (up) Stone/gettyimages; 154 (lo) Courtesy Susan S. Blair; 154-5 Mark Gamba/CORBIS; 156 NASA/Photo Researchers; 157 Stone/gettyimages; 158 (le) Dwight R. Kuhn; 158 (rt) Peter Arnold Inc.; 158-9 CORBIS; 160 Bill Curtsinger; 160-1 Grant Heilman Photography; 161 Runk/Schoenberger/Grant Heilman Photography; 162 The Seattle Times/Liaison Agency/gettyimages; 162-3 Robert W. Tope, Natural Science Illustrations; 163 Andy Levin/Photo Researchers; 163 (inset) Gerald Zanetti; 164-5 AP Photo/Michael Scates; 166-7 Stone/gettyimages; 168 (le) Richard Thompson; 168 (rt) Gary W. Carter/CORBIS; 169 Photo used with permission: plastics.org

The Chemistry of Life

170-1 Lennart Nilsson; 172 (le) NIH/Photo Researchers; 172 (rt) Theophilus Britt Griswold; 172-3 The Image Bank/gettyimages; 174 Rosemarie D'Alba; 174-5 Carl W. Rohrig; 176 (up) Photodisc Green/gettyimages; 176 (lo) The Image Bank/gettyimages; 177 Sally J. Bensusen Visual Science Studio; 178 (up) Brian Hagiwara/gettyimages; 178 (lo) David Aubrey/CORBIS; 179 Adam Pretty/gettyimages; 180 Allsport Concepts/gettyimages; 181 Bill Hatcher; 182 David Madison/Duomo Photography Inc; 183 Nigel Marple/gettyimages; 184-5 The Image Bank/gettyimages; 186 Richard Thompson; 186-7 David M. Phillips/Photo Researchers; 188 Gerry Davis/Phototake; 189 Wood Ronsaville Harlin, Inc.; 190 Richard Thompson; 190-1 Carl W. Rohrig; 191 Stone/gettyimages; 192 Kirk Moldoff; 192-3 (up) CORBIS; 192-3 (lo) Kirk Moldoff; 193 Richard Thompson; 194 Strauss/Curtis/CORBIS; 194-5 LWA-Dann Tardif/CORBIS; 196 Barry Rosenthal/Taxi/gettyimages; 196-7 Stephen Simpson/Taxi/gettyimages; 197 John W. Karapelou/Phototake; 198-9 Wood Ronsaville Harlin, Inc.; 198-9 Wood Ronsaville Harlin, Inc.; 200 A. Liepins/Photo Researchers; 200-1 Stem Jems/Photo Researchers; 202 Robert J. Demarest; 203 Jose Luis Pelaez Inc./CORBIS; 203 (inset) Dr. Scott Camazine; 204 (both) Science Photo Library/Photo Researchers; 205 Lennart Nilsson; 206-7 Lennart Nilsson; 208 O. Louis Mazzatenta; 208-9 Clouds Hill Imaging Ltd/CORBIS; 210 Teri J. McDermott/Phototake; 210-1 Alex Wong/gettyimages; 211 Culver Pictures; 212 Richard Thompson; 212-3 Dr. Jason Weisfeld; 214 NIH/Scott Camazine/Photo Researchers; 215 Kirk Moldoff; 216-7 Carl W. Rohrig; 218 Carolina Biological Supply Co/Phototake; 218-9 CORBIS; 219 Teri J. McDermott/Phototake; 220 A. Liepins/Photo Researchers; 220-1 Sally J. Bensusen Visual Science Studio; 222-3 CORBIS; 224 C. Briscoe/Photo Researchers; 224-5 Stephen Dunn/gettyimages; 226 Daniel J. Edelman Inc Public Relations; 226-7 (all) Kimberly Martens & Scott Barrows CMI, FAMI; 228 Teri J. McDermott/Phototake; 228-9 (up) Charles O'Rear/CORBIS; 228-9 (lo) Randy Faris/CORBIS; 230 Joe McNally; 231 Photo Researchers; 232 The Image Bank/gettyimages; 232-3 Scott Barrow; 233 Robb Kendrick Photography; 234-5 (all) Wood Ronsaville Harlin, Inc.; 236 Scott Barrow; 237 (up) Richard Thompson; 237 (lo) B. Barley/ SuperStock; 238 Theophilus Britt Griswold; 238-9 Comet Photography Inc.; 240 Scott Barrow; 241 Carter Blair; 242-3 Tek Image/Photo Researchers; 243 (inset) Dennis Kunkel/Phototake; 244 Richard Thompson; 244-5 Wood Ronsaville Harlin, Inc.; 246 Richard Thompson; 246-7 Joe McNally; 247 (up le) Ralph C. Eagle Jr./Photo Researchers; 247 (up rt) Teri J. McDermott/Phototake; 248 (both) Theophilus Britt Griswold; 249 Massimo Listri/CORBIS; 250-1 Chistos Kalohoridis/CORBIS; 252 Wood Ronsaville Harlin, Inc.; 252-3 The Image Bank/gettyimages; 254 Wood Ronsaville Harlin, Inc.; 255 Taxi/gettyimages; 256 Michael Keller/CORBIS; 256-7 Stuart Franklin/gettyimages; 258-9 Lindsay Hebberd; 260 Brand X Pictures/gettyimages; 261 UPI/Bettman/CORBIS; 262 (up) FoodPix/gettyimages; 262 (lo) Digital Vision/gettyimages; 263 Stone/gettyimages; 264 Richard Thompson; 264-5 SuperStock; 266 Scott Thorn Barrows; 266-7 Peter Angelo Simon/Phototake